From ASICs to SOCs

A Practical Approach

Prentice Hall Modern Semiconductor Design Series

James R. Armstrong and F. Gail Gray
   VHDL Design Representation and Synthesis

Mark Gordon Arnold
   Verilog Digital Computer Design: Algorithms into Hardware

Jayaram Bhasker
   A VHDL Primer, Third Edition

Eric Bogatin
   Signal Integrity: Simplified

Douglas Brooks
   Signal Integrity Issues and Printed Circuit Board Design

Kanad Chakraborty and Pinaki Mazumder
   Fault-Tolerance and Reliability Techniques for High-Density Random-Access Memories

Ken Coffman
   Real World FPGA Design with Verilog

Alfred Crouch
   Design-for-Test for Digital IC's and Embedded Core Systems

Daniel P. Foty
   MOSFET Modeling with SPICE: Principles and Practice

Nigel Horspool and Peter Gorman
   The ASIC Handbook

Howard Johnson and Martin Graham
   High-Speed Digital Design: A Handbook of Black Magic

Howard Johnson and Martin Graham
   High-Speed Signal Propagation: Advanced Black Magic

Pinaki Mazumder and Elizabeth Rudnick
  Genetic Algorithms for VLSI Design, Layout, and Test Automation

Farzad Nekoogar and Faranak Nekoogar
  From ASICs to SOCs: A Practical Approach

Farzad Nekoogar
  Timing Verification of Application-Specific Integrated Circuits (ASICs)

David Pellerin and Douglas Taylor
  VHDL Made Easy!

Samir S. Rofail and Kiat-Seng Yeo
  Low-Voltage Low-Power Digital BiCMOS Circuits: Circuit Design, Comparative Study, and Sensitivity Analysis

Frank Scarpino
  VHDL and AHDL Digital System Implementation

Wayne Wolf
  Modern VLSI Design: System-on-Chip Design, Third Edition

Kiat-Seng Yeo, Samir S. Rofail, and Wang-Ling Goh
  CMOS/BiCMOS ULSI: Low Voltage, Low Power

Brian Young
  Digital Signal Integrity: Modeling and Simulation with Interconnects and Packages

Bob Zeidman
  Verilog Designer's Library

About Prentice Hall Professional Technical Reference

With origins reaching back to the industry's first computer science publishing program in the 1960s, and formally launched as its own imprint in 1986, Prentice Hall Professional Technical Reference (PH PTR) has developed into the leading provider of technical books in the world today. Our editors now publish over 200 books annually, authored by leaders in the fields of computing, engineering, and business.

Our roots are firmly planted in the soil that gave rise to the technical revolution. Our bookshelf contains many of the industry's computing and engineering classics: Kernighan and Ritchie's *C Programming Language*, Nemeth's *UNIX System Adminstration Handbook*, Horstmann's *Core Java*, and Johnson's *High-Speed Digital Design*.

PH PTR acknowledges its auspicious beginnings while it looks to the future for inspiration. We continue to evolve and break new ground in publishing by providing today's professionals with tomorrow's solutions.

From ASICs to SOCs

A Practical Approach

Farzad Nekoogar
Faranak Nekoogar

PRENTICE HALL
Professional Technical Reference
Upper Saddle River, NJ 07458
www.phptr.com

Library of Congress Cataloging-in-Publication Data

```
Nekoogar, Farzad.
    From ASICS to SOCs: a practical approach / Farzad Nekoogar,
  Faranak Nekoogar.
        p. cm. — (Prentice Hall modern semiconductor design series)
     Includes bibliographical references and index.
     ISBN 0-13-033857-5 (case)
     1. Application specific integrated circuits.    2. Systems on a
  chip.    I. Nekoogar, Faranak.    II. Title.    III. Series.
  TK7874.6.N43   2003
  621.3815—dc21
                                                      2003043862
```

Editorial/production supervision: *BooksCraft, Inc.*
Cover design director: *Jerry Votta*
Cover designer: *Nina Scuderi*
Art director: *Gail Cocker-Bogusz*
Manufacturing buyer: *Maura Zaldivar*
Publisher: *Bernard Goodwin*
Editorial assistant: *Michelle Vincenti*
Marketing manager: *Dan DePasquale*
Full-service production manager: *Anne R. Garcia*

© 2003 by Pearson Education, Inc.
Publishing as Prentice Hall Professional Technical Reference
Upper Saddle River, New Jersey 07458

Prentice Hall books are widely used by corporations and government agencies for training, marketing, and resale.

Prentice Hall PTR offers excellent discounts on this book when ordered in quantity for bulk purchases or special sales. For more information, please contact:
   U.S. Corporate and Government Sales
   1-800-382-3419
   corpsales@pearsontechgroup.com

For sales outside of the U.S., please contact:
   International Sales
   1-317-581-3793
   international@pearsontechgroup.com

Company and product names mentioned herein are the trademarks or registered trademarks of their respective owners.
All rights reserved. No part of this book may be reproduced, in any form or by any means, without permission in writing from the publisher.

Printed in the United States of America
1st Printing

ISBN 0-13-033857-5

Pearson Education LTD.
Pearson Education Australia PTY, Limited
Pearson Education Singapore, Pte. Ltd.
Pearson Education North Asia Ltd.
Pearson Education Canada, Ltd.
Pearson Educación de Mexico, S.A. de C.V.
Pearson Education—Japan
Pearson Education Malaysia, Pte. Ltd.

To our older brother Farhad, who opened the gate to great opportunities for both of us.
—Farzad and Faranak

Contents

List of Abbreviations xiii

Preface xvii

Acknowledgments xxi

1 Introduction 1
   1.1 Introduction 1
   1.2 Voice Over IP SOC 2
   1.3 Intellectual Property 7
   1.4 SOC Design Challenges 12
   1.5 Design Methodology 16
   1.6 Summary 18
   1.7 References 20

2 Overview of ASICs 21
   2.1 Introduction 21
   2.2 Methodology and Design Flow 25
   2.3 FPGA to ASIC Conversion 32
   2.4 Verification 34

	2.5 Summary	40
	2.6 References	41
3	**SOC Design and Verification**	43
	3.1 Introduction	43
	3.2 Design for Integration	44
	3.3 SOC Verification	47
	3.4 Set-Top-Box SOC	56
	3.5 Set-Top-Box SOC Example	57
	3.6 Summary	79
	3.7 References	79
4	**Physical Design**	81
	4.1 Introduction	81
	4.2 Overview of Physical Design Flow	82
	4.3 Some Tips and Guidelines for Physical Design	87
	4.4 Modern Physical Design Techniques	93
	4.5 Summary	108
	4.6 References	109
5	**Low-Power Design**	111
	5.1 Introduction	111
	5.2 Power Dissipation	112
	5.3 Low-Power Design Techniques and Methodologies	117
	5.4 Low-Power Design Tools	140
	5.5 Tips and Guidelines for Low-Power Design	145
	5.6 Summary	146
	5.7 References	147
A	**Low-Power Design Tools**	151
	PowerTheater	151
	PowerTheater Analyst	153
	PowerTheater Designer	155
B	**Open Core Protocol (OCP)**	159
	Highlights	160
	Capabilities	160

　　　　Advantages 161
　　　　Key Features 162

C　Phase-Locked Loops (PLLs) 165
　　　　PLL Basics 165
　　　　PLL Ideal Behavior 166
　　　　PLL Errors 168

　　Glossary 173

　　Index 183

List of Abbreviations

AAL1	ATM Adaptation Layer 1
AAL2	ATM Adaptation Layer 2
ABV	Assertion-Based Verification
AC	Alternating Current
ADC	Analog-to-Digital Converter
ADPCM	Adaptive Differential Pulse Code Modulation
ASIC	Application-Specific Integrated Circuit
ATM	Asynchronous Transfer Mode
ATPG	Automatic Test Pattern Generation
BFM	Bus Functional Model
BGA	Ball Grid Array
BIST	Built-In Self Test
CAD	Computer Aided Design
CELP	Code Excited Linear Predictive
CMOS	Complementary Metal Oxide Semiconductor
CODEC	COder/DECoder
CPCI	Compact Peripheral Component Interconnect
CTV	Cable TV
CVS	Concurrent Versions System
DAC	Digital-to-Analog Converter

DC	Direct Current
DDR	Double Data Rate
DDS	Digital Data Service
DFT	Design For Test
DIP	Dual In-Line Package
DLL	Digital Link Layer
DMA	Direct Memory Access
DRAM	Dynamic Random Access Memory
DRC	Design Rule Check
DSL	Digital Subscriber Line
DSM	Deep Sub-Micron
DSP	Digital Signal Processing/ Digital Signal Processor
DTMF	Dual-Tone Multi Frequency
DUT	Design Under Test
ECO	Engineering Change Orders
EDA	Electronic Design Automation
EDIF	Electronic Design Interchange Format
ERC	Electrical Rule Check
ESD	Electrostatic Discharge
FIFO	First-In First-Out
FPGA	Field Programmable Gate Array
FSM	Finite State Machine
GND	Ground
GPS	Global Positioning System
HDL	Hardware Description Language
HLB	Hierarchical Layout Block
HW/SW	Hardware/Software
ICs	Integrated Circuits
ILM	Interface Logic Models
IP	Intellectual Property
IP	Internet Protocol
IPO	In Place Optimization
IR	commonly refers to voltage drop from $V = IR$
ISDN	Integrated Services Digital Network
ITU	International Telecommunication Union
JTAG	Joint Test Action Group

List of Abbreviations

K-maps	Karnaugh maps
LEC	Line Echo Canceller
LVS	Layout Versus Schematic
MAC	Media Access Control
MII	Media Independent Interface
MPEG	Moving Picture Experts Group
MPU	MicroProcessor Unit
MVIP	Multi Vendor Integration Protocol
NMOS	N-channel Metal-Oxide-Semiconductor
NRE	Non-Recurring Engineering
OCB	On-Chip Buses
OCP	Open Core Protocol
OIF	Optical Internetworking Forum
PCB	Printed Circuit Board
PCI	Peripheral Component Interconnect
PCM	Pulse Code Modulation
PGA	Pin Grid Array
PIP	Picture In Picture
PLL	Phase Locked Loops
PMOS	P-channel Metal-Oxide-Semiconductor
PSTN	Public Switched Telephone Network
PVT	Process, Voltage, and Temperature
QFP	Quad Flat Pack
QAM	Quadrature Amplitude Modulation
QPSK	Quadrature Phase Shift Keying
RCS	Revision Control System
RGB	Red-Green-Blue
RISC	Reduced Instruction Set Computer
RMII	Reduced Media Independent Interface
RT	Register Transfer
RTL	Register Transfer Level
SB	SiliconBackplane
SCSA	Signal Computing System Architecture
SDC	SDRAM Controller
SDF	Standard Delay Format
SDRAM	Synchronous Dynamic Random Access Memory

Serdes	Serializer/Deserializer
SFI	Serdes-to-Framer Interface
SI	Signal Integrity
SOC	System On a Chip
SOP	Small Outline Package
SPI-4P2	System Packet Interface Level 4 Phase 2
STA	Static Timing Analysis
STB	Set Top Box
STV	Satellite TV
TAT	Turn Around Time
TCP	Transfer Control Protocol
TDM	Time Division Multiplexing
TSI	Time Slot Interchange
TTM	Time To Market
UDP	User Datagram Protocol
USB	Universal Serial Bus
UTOPIA	Universal Test Operation PHY Interface for ATM
VAD	Voice Activity Detector
VC	Virtual Components
VCI	Virtual Component Interface
VHDL	VHSIC (Very high-speed integrated circuit) Hardware Description Language
VOCODER	Voice CODER
VoIP	Voice over IP
VoN	Voice over Network
VSIA	Virtual Socket Interface Alliance
WAN	Wide Area Network
WLM	Wire Load Models
xDSL	Digital Subscriber Line
XNF	Xilinx Netlist Format

Preface

The term SOC (system-on-a-chip) has been used in the electronic industry over the last few years. However, there are still a lot of misconceptions associated with this term. A good number of practicing engineers don't really understand the differences between ASICs and SOCs. The fact that the same EDA tools are used for both ASICs and SOCs design and verification doesn't help to reduce the misconceptions.

This book describes the practical aspects of ASIC and SOC design and verification. It reflects the current issues facing ASIC/SOC designers.

The following items characterize the book:

- It deals with everyday issues that ASIC/SOC designers have to face as opposed to generic textbook examples covered in other books.
- It emphasizes principles and techniques as opposed to specific tools. Once the designers understand the underlying principles of practical design, they can apply them with various tools.

- ☞ FPGAs will not be covered in this book. However, in Chapter 2 we cover a short section on FPGA to ASIC conversion. Earlier books have covered design and verification of FPGAs adequately.
- ☞ It provides tips and guidelines for front-end and back-end designs.
- ☞ Modern physical design techniques are covered.
- ☞ Low-power design techniques and methodologies are explored for both ASICs and SOCs.

This book is to be used for self-study by practicing engineers. Design and verification engineers who are working with ASICs and SOCs will find the book very useful. Upper-level undergraduate and graduate students in electrical engineering can use it as a reference book in courses in logic and chip design and related topics.

The material covered in the book requires understanding of EDA tools as well as front-end and back-end processes in chip design. An initial course in logic design is required.

The book is organized in the following fashion.

In Chapter 1 we introduce the goals of this manuscript. The differences between ASICs and SOCs are introduced. The concept of Intellectual Property (IP) is covered as well as an overview of design methodologies.

SOC design challenges such as integration of IPs are also covered.

A gateway VOIP (Voice Over IP) SOC example is given in this chapter.

Chapter 2 covers an overview of ASIC design concepts, methodology, and front-end design flow. Useful guidelines for hierarchical design methodology are presented such as placement-based synthesis and interface logic models. Some key questions that ASIC designers should consider when designing ASICs are presented. FPGA to ASIC conversion is covered in Section 2.3. An overview of verification and Design for Test (DFT) techniques are also presented in this chapter.

Chapter 3 continues with the VOIP SOC example from Chapter 1. Design for integration is covered in Section 3.2. Section 3.3 covers SOC verification planning guidelines such as resource planning and regression planning. Automation and IP verification are also covered in Section 3.3. This chapter ends with a detailed design example of a Set-Top Box (STB).

Chapter 4 covers an overview of the physical design flow. Some tips and guidelines for physical design are given such as logical vs. physical hierarchy, multiple placements and routing, and non-routable congested areas.

Two examples of modern physical-design techniques are presented in Section 4.4. These methods each overcome the problems associated with traditional physical design techniques.

In Chapter 5 we present low-power design techniques. In this chapter, sources of power dissipation in CMOS devices are discussed. Several methods of power optimization at various levels of abstraction for ASICS and SOCs are explained. These techniques include: algorithm-level optimization, architecture-level optimization, RT-level optimization, and gate-level optimization. Appendix A should be used in conjunction with this chapter.

Appendix A summarizes EDA low-power design tools from Sequence Design, Inc.

Appendix B gives an overview of OCP (Open Core Protocol) that is used as a core interface standard for IP integration.

Appendix C gives an introduction to Phase-Locked Loops which are widely used in almost all ASICs and SOCs.

Acknowledgments

We are indebted to Professor Wayne Wolf of the electrical engineering department at Princeton University and Richard Rubinstein for their detailed review of the manuscript, constructive criticism, and suggestions of information to be added.

In addition we would like to thank the following people and companies:

- The staff of Prentice Hall, especially Bernard Goodwin, for his support of this project
- Ken Schmidt for reviewing the chapter on low power
- Ron Sailors for reviewing parts of the book
- Farshid Tabrizi and Munir Ahmed of Ammocore Technology Inc.
- Michel Courtoy, vice president of marketing at Silicon Prespective, Inc. (A Cadence Company)
- Plato Design Systems (A Cadence Company)
- Fujitsu Microelectronics of America
- Sequence Design, Inc.

- OCP-IP association
- The staff of BooksCraft, Inc., for their help in producing the book

Farzad Nekoogar
Faranak Nekoogar

CHAPTER 1

Introduction

1.1 INTRODUCTION

The ASIC (Application Specific Integrated Circuit) and SOC (System on a Chip) abbreviations are used every day in the integrated circuit design industry. However, there are still a lot of ambiguities when differentiating SOCs from traditional ASICs. Some designers define SOCs as complex integrated circuits with more than one on-chip processor. Many use the term when describing ICs that have more than 10 million gates plus on-chip processors. Still others define it as ICs that contain soft and hard functional blocks as well as digital and analog components. Let's give our own definition here.

An SOC is a system on an IC that integrates software and hardware Intellectual Property (IP) using more than one design methodology for the purpose of defining the functionality and behavior of the proposed system. In most cases, the designed system is application specific. Typical applications can be found in the consumer, networking, communications, and other segments of the electronics industry. Voice over Internet Protocol (VoIP) is a good example of an emerging market where SOCs are widely designed. Figure 1.1 shows an example of a typical gateway VoIP system-on-a-chip diagram.

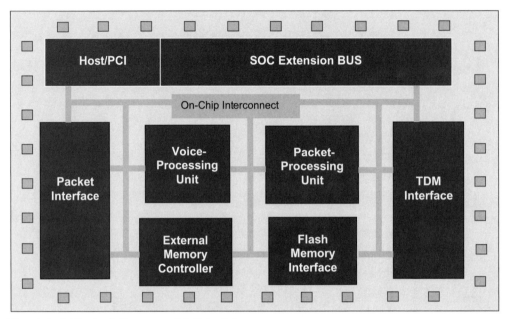

Fig. 1.1 A Typical Gateway SOC Architecture

1.2 Voice Over IP SOC

A gateway VoIP SOC is a device used for functions such as vocoders, echo cancellation, data/fax modems, and VoIP protocols. Currently, there are a number of these devices available from several vendors; typically these devices differ from each other by the type of functions and voice-processing algorithms they support.

In this example, we define the major blocks required to support carrier-class voice processing. The SOC can vary depending on the particular I/O and voice-processing requirements of the mediation gateway architecture. Major units for this SOC are as follows.

Host/PCI

The host interface is for control, code download, monitoring, and in some cases data transport. This host interface could be either a microprocessor-specific interface or a generic system-bus interface such as PCI.

- **Microprocessor Interface** A synchronous processor interface, such as a 32-bit synchronous Motorola 68000 or Intel 960 style interface operating at 33MHz with interrupt support, allows the SOC to interface to most processors with minimal glue logic. This interface usually supports multiplexed data and addresses to reduce the number of I/Os on the SOC. The SOC also supports interrupt generation in order to notify the CPU of external events.
- **PCI Interface** The SOC may have a PCI-compliant interface for communication with external processors and resources. The PCI interface would support bus Target (Slave) and Initiator (Master) functions and DMA, but would not require an arbiter. This interface also provides access to shared memory.

External Memory Controller

The external memory controller supports industry-standard inexpensive fast memory such as SDRAM. This memory is used to store code and data for processing elements within the SOC. Depending on the actual SOC architecture and fabrication process, the memory interface could require support for one 32-bit SDRAM module, two 16-bit modules operating at up to 133MHz.

Flash Memory Interface

A standard parallel flash port for access to boot programs, configuration data, and programs is available and accessible upon system reset.

Packet Interface

The packet interface can be Ethernet or Utopia.

- **Ethernet** A standard 10/100BT Ethernet MII or RMII interface may be useful in cases where both compression and packetization are performed in the SOC. In such architectures, IP packets may be transported within a system using Ethernet as the physical transport layer.

- **Utopia** An industry standard, Utopia level 2 interface is useful for interfacing to system fabrics that use ATM as a physical transport. This interface supports connections to ATM 155Mbit/s physical-layer interfaces.

TDM Interface

The TDM interface is the downstream interface to PSTN TDM streams. These are uncompressed voice channels of 64Kbit/s A-LAW/µ-LAW voice that is delivered to the SOC for compression and forwarding to the packet network. The SOC interfaces directly with legacy TDM device interfaces such as the ECTF H.100/H.110 standard serial interface.

- **ECTF H.100/H.110** H.100/H.110 is a standard TDM interface for legacy telephony equipment. H.100/H.110 allows the transport of up to 4096 simplex channels of voice or data on one connector or ribbon cable. This voice traffic may come from a WAN interface board, chip, or any other voice-processing device in the carrier systems described above. H.100 defines a mezzanine connection that can interface to other H.100 devices or to legacy MVIP/SCSA devices.

SOC Extension Bus

The SOC extension bus is required to load balance the system and to provide a unified host interface for access.

Voice/Tone Processing Unit

The voice/tone processing unit consists of multiple DSP cores that perform the following functions:

- Code excited linear prediction (CELP)
- Pulse code modulation (PCM)
- Echo cancellation
- Silence suppression
- Voice activity detector (VAD)
- Tone detection/generation
- Dual-tone multifrequency (DTMF)

Packet Processing Unit

The packet-processing unit consists of several packet processors that process the voice and signaling packets that are ready for transmission. This unit performs the following functions:

- ATM Adaptation Layer 1 (AAL1)
- ATM Adaptation Layer 2 (AAL2)
- User Datagram Protocol (UDP)
- Transfer Control Protocol (TCP)

We will spend more time on this gateway SOC in Chapter 3. Let's look at another SOC example. Figure 1.2 shows an overview diagram of a set-top-box (STB) SOC.

The major blocks in Figure 1.2 and their functions are listed below:

- Video processing unit (MPEG-2 codec)
- Digital signal processing (DSP) for AC3 audio processing
- CPU for control and transport of streams
- Modulation unit such as quadrature phase shift keying (QPSK) for satellite and quadrature amplitude modulation (QAM) for cable inputs

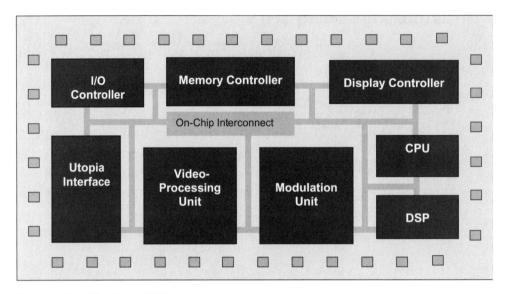

Fig. 1.2 Set-Top-Box SOC

- Utopia for cable modem interface
- Memory controller such as SDRAM controller
- I/O controller
- Display controller

A more detailed example of an STB is presented in Section 3.4.
In many SOC designs, you will find the following characteristics:

- Hierarchical architecture
- Hierarchical methods for physical design (placement and routing) and timing analysis
- On-chip interconnect
- Standard core-to-core communication protocols
- Hardware/Software codesign/verification
- Reusable infrastructure

Before we go further on SOC design, we need to introduce the concept of an IP.

1.3 INTELLECTUAL PROPERTY

In today's rapidly growing IC technology, the number of gates per chip can reach several millions, exceeding Moore's law: "The capacity of electronic circuits doubles every 18 months." To overcome the design gap generated by such fast-growing capacity and lack of available manpower, reuse of the existing designs becomes a vital concept in design methodology. IC designers typically use predesigned modules to avoid reinventing the wheel for every new product. Utilizing the predesigned modules accelerates the development of new products to meet today's time-to-market challenges. By practicing design-reuse techniques—that is, using blocks that have been designed, verified, and used previously—various blocks of a large ASIC/SOC can be assembled quite rapidly. Another advantage of reusing existing blocks is to reduce the possibility of failure based on design and verification of a block for the first time. These predesigned modules are commonly called Intellectual Property (IP) cores or Virtual Components (VC).

Designing an IP block generally requires greater effort and higher cost. However, due to its reusable architecture, once an IP is designed and verified, its reuse in future designs saves significant time and effort in the long run. Designers can either outsource these reusable blocks from third-party IP vendors or design them inhouse. Figure 1.3 represents an approximation of the amount of resources used in several designs with and without utilizing the design-reuse techniques.

As shown in Figure 1.3, the time and cost to design the first reusable block are higher than those for the design without reusability. However, as the number of usages increases, the time-saving and cost-saving benefits become apparent.

Licensing the IP cores from IP provider companies has become more popular in the electronic industry than designing inhouse reusable blocks for the following reasons:

1. Lack of expertise in designing application-specific reusable building blocks.

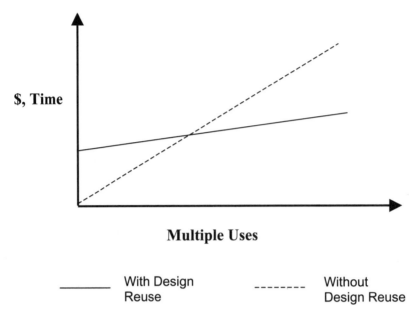

Fig. 1.3 Resources versus Number of Uses

2. Savings in time and cost to produce more complex designs when using third-party IP cores.
3. Ease of integration for available IP cores into more complicated systems.
4. Commercially available IP cores are preverified and reduce the design risk.
5. Significant improvement to the product design cycle.

Intellectual Property Categories

To provide various levels of flexibility for reuse and optimization, IP cores are classified into three distinct categories: hard, soft, and firm.

Hard IP cores consist of hard layouts using particular physical design libraries and are delivered in masked-level designed blocks (GDSII format). These cores offer optimized implementation and the highest performance for their chosen physical library. The integration of hard IP cores is quite simple and the core can be dropped into an SOC physical design with minor integration effort. However, hard cores are technology dependent and provide minimum flexibility and portability in reconfiguration and integration across multiple designs and technologies.

Soft IP cores are delivered as RTL VHDL/Verilog code to provide functional descriptions of IPs. These cores offer maximum flexibility and reconfigurability to match the requirements of a specific design application. Although soft cores provide the maximum flexibility for changing their features, they must be synthesized, optimized, and verified by their user before integration into designs. Some of these tasks could be performed by IP providers; however, it's not possible for the provider to support all the potential libraries. Therefore, the quality of a soft IP is highly dependent on the effort needed in the IP integration stage of SOC design.

Firm IP cores bring the best of both worlds and balance the high performance and optimization properties of hard IPs with the flexibility of soft IPs. These cores are delivered in the form of targeted netlists to specific physical libraries after going through synthesis without performing the physical layout. Figure 1.4 represents the role of firm IP cores in ASIC design flow.

In Figure 1.4, the tasks in shaded boxes can be covered by Firm IP and as a result accelerate the design flow. Table 1.1 provides a brief comparison of different IP formats.

Table 1.2 provides a collection of some of the deliverable items for different IP formats.

Table 1.1 Comparison of Different Intellectual Property Formats

IP Format	Representation	Optimization	Technology	Reusability
Hard	GDSII	Very High	Technology Dependent	Low
Soft	RTL	Low	Technology Independent	Very High
Firm	Targeted Netlist	High	Technology Generic	High

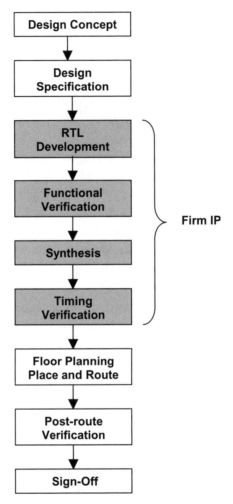

Fig. 1.4 ASIC Design Flow

Guidelines for Outsourcing IP

Although licensing IP can greatly enhance project design cycles, it can also hurt project schedules if the following are not carefully considered when selecting an IP vendor.

- ☞ Outsource IPs from a well-known IP provider with large customer base and great track record. Customer testimonials of integrating a specific IP from a third-party vendor represent

Table 1.2 Some of the Deliverables for Various IP Formats

Deliverables	Hard IP	Soft IP	Firm IP
HDL RTL code		•	
HDL targeted netlist			•
GDSII file	•		
Functional verification testbenches	•	•	•
Bus functional models	•	•	•
Floor planning models	•		
Synthesis and timing models	•	•	•
Full documentation	•	•	•

the best way of ensuring that the IP works in the integration process.

☞ Evaluate the IP functionality using demos and executable models before purchasing. Hardware demonstrations by IP providers are another way of ensuring that IP blocks are functional in silicon. Access to executable models allows you to change different parameters and make sure the IP provides functional results that you expect for your design.

☞ Ask for a full verification test environment. A full verification environment provides a set of models for different stimuli to verify the IP functionality and makes the overall chip verification less complicated.

☞ IPs should be accompanied by detailed documentation, such as datasheet, databook, user's guide, application notes, etc. Proper documentation offers valuable information on timing, interface definition, and different configurations for specific applications.

☞ Allocate a certain period of time to become familiar with the interfaces and functionality of the outsourced IP. It is quite common that IP interfaces do not match the rest of the system interface causing additional work to be done in the integration process. This could change the project schedule if the additional integration time is not included in the project timeline.

☞ Make an agreement with the IP provider for technical support during the integration process. There are many instances when an IP has to be customized for a specific design at the integration time and only the IP provider is able to perform these modifications. Therefore, it is necessary to have the IP provider's support through the integration process.

We will cover more on IP verification and integration in Chapter 3. Table 1.3 shows several examples of Silicon IPs.

Table 1.3 Examples of IPs

Category	Intellectual Property
Processor	ARM7, ARM9, and ARM10, ARC
Application-Specific DSP	ADPCM, CELP, MPEG-2, MPEG-4, Turbo Code, Viterbi, Reed Solomon, AES
Mixed Signal	ADCs, DACs, Audio Codecs, PLLs, OpAmps, Analog MUX
I/Os	PCI, USB, 1394, 1284, E-IDE, IRDA
Miscellaneous	UARTs, DRAM Controller, Timers, Interrupt Controller, DMA Controller, SDRAM Controller, Flash Controller, Ethernet 10/100 MAC

1.4 SOC DESIGN CHALLENGES

Why does it take longer to design SOCs compared to traditional ASICs? To answer this question, we must examine factors influencing the degree of difficulty and Turn Around Time (TAT) for designing ASICs and SOCs. Usually for an ASIC, the following factors influence TAT:

☞ Frequency of the design
☞ Number of clock domains

1.4 SOC DESIGN CHALLENGES

☞ Number of gates
☞ Density
☞ Number of blocks

Another factor that influences TAT for SOCs is system integration (mainly integrating different silicon IPs on the same IC) that is one of the key factors in TAT. In a typical SOC, you deal with complex data flows and multiple cores such as CPUs, DSPs, DMA, and peripherals. Therefore, resource sharing becomes an issue. Figure 1.5 shows a bus-based approach to integration. Here, the architecture is tightly coupled, which is advantageous for performance, area,

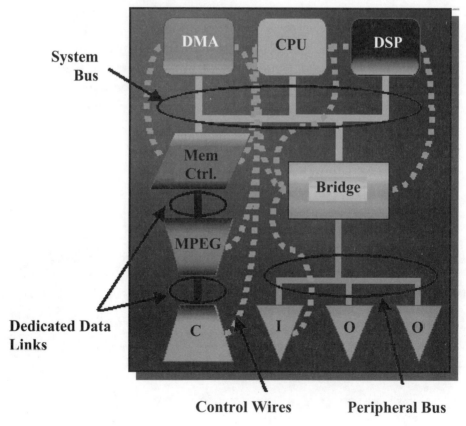

Fig. 1.5 A Traditional SOC Architecture (Copyright 2002, Sonics, Inc.)

and efficiency. However, communication between IPs becomes very complicated.

Let's examine this approach, as it is common practice among chip architects and designers. Here the CPU, DMA, and the DSP engine all share the same bus (the CPU or the system bus). Also, there are dedicated data links and a lot of control wires between blocks. Additionally, there are peripheral buses between subsystems. As a result, there is excessive interdependency between blocks and a lot of wires in the chip. Therefore, verification, test, and physical design all become difficult to fulfill.

A solution to this system integration is to use an intelligent, on-chip interconnect that unifies all the traffic into a single entity. An example of this is Sonics' SMART Interconnect SiliconBackplane MicroNetwork.

A MicroNetwork is a heterogeneous, integrated network that unifies, decouples, and manages all of the communication between processors, memories, and input/output devices. Figure 1.6 shows an SOC design using MicroNetwork architecture. An example of a MicroNetwork is Sonics' SiliconBackplane, which guarantees end-to-end performance by managing all communications among IP

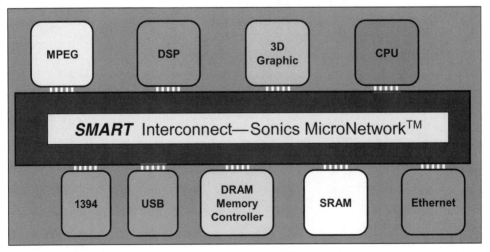

Fig. 1.6 Sonics' SiliconBackplane Used in SOC Design Architecture (Copyright 2002, Sonics, Inc.)

cores, as well as ensuring high-speed access to the shared memories common in typical SOC designs.

SiliconBackplane uses a standard core interface known as the Open Core Protocol (OCP), which delivers the first openly licensed, core-centric protocol. OCP comprehensively fulfills system-level integration requirements. The OCP defines a comprehensive, bus-independent, high-performance, and configurable interface between IP cores and on-chip communication subsystems. OCP is a functional superset of the Virtual Socket Interface (VSI) Alliance virtual-component-interface (VCI) specification, and enables SOC designers and semiconductor IP developers to prepare their cores for plug-and-play integration using Sonics' SiliconBackplane. Appendix B provides more information on OCP.

An SOC designer can optimize the design under development by optimizing the SiliconBackplane using a development environment developed by Sonics. Configuration and tuning parameters can be efficiently selected to optimize the SiliconBackplane and, as a result, to optimize the SOC design. The development environment consists of tools to wrap and package IP cores for integration as well as an automated basic configuration of the SiliconBackplane, and stimulus/performance analysis tools for successively refining SOCs.

When compared to a traditional CPU bus, an on-chip interconnect such as Sonics SiliconBackplane has the following advantages:

- Higher efficiency
- Flexible configuration
- Guaranteed bandwidth and latency
- Integrated arbitration

Design verification is another key challenge in designing SOCs. Verification has to happen at all levels of hierarchy, such as core/IP level, interface, and chip level. The integration of several cores on a single chip brings with it new challenges to the testing methodology even when the individual cores have design for test (DFT) already built in. The cores may have different types of testability: scan, built-in self-test (BIST), and functional. The integrator of the cores must decide on a coherent test style from the outset and choose the cores accordingly. This, in turn, implies that the integrator has access to a number of IP providers and he or she has established an acceptance criterion for cores.

Chapter 3 covers the verification of cores and SOCs in more detail.

1.5 DESIGN METHODOLOGY

A good design methodology for ASICs and SOCs consists of a set of defined design flows for both front and back ends as well as tool integration and task automation. Let's start with the design flow. Figure 1.4 showed a typical top-down ASIC design flow. The flow can be divided into the following major parts: design entry, design implementation, design verification, physical design, and IC production. A more detailed front-end flow diagram is shown in Figure 1.7. Let's look at the steps involved in this flow.

The designer develops the RTL code that implements the functional specification. Chip designers should follow any coding guidelines provided by ASIC vendors.

Simulations at the register-transfer (RT) level should be thorough because this is really the only place where correct function can be verified efficiently. Simulations at the gate level are much too slow to be complete and static timing analysis (STA) does not verify functionality, only timing.

The synthesis tool generates both forward and backward annotation files. The forward annotation provides constraints to timing-driven layout tools while the back-annotated files provide delay information to either a simulator for gate-level simulations or a static timing analyzer.

The designer is responsible for verifying the synthesized gates for functional correctness and for estimated performance. Whether the verification is done with a simulator or a static timing analyzer, the wire loads are only estimates. The gate delays come from the

1.5 Design Methodology

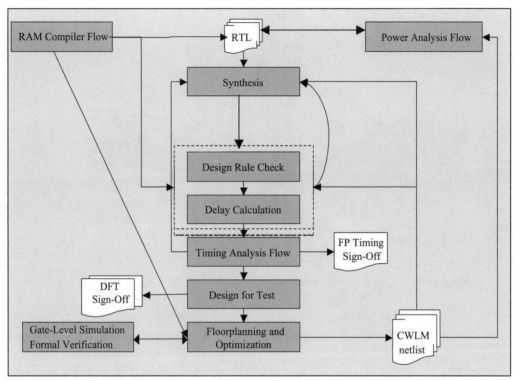

Fig. 1.7 A Front-End ASIC Design Flow (Printed with permission of Fujitsu Microelectronics America, Inc.)

technology library and are accurate. The delays are provided from the synthesis tool via a standard delay format (SDF) file.

Floorplanning takes information from the synthesis step to group the cells to meet the timing performance. It feeds back more accurate wire-load models to the synthesis tool and it provides the framework for place and route.

Figure 1.8 shows a spiral design flow. This type of flow is becoming popular with SOC designers for the front end. Here, the designers work simultaneously on each phase of the design until the design is gradually completed.

Once you finish the front-end work and generate a gate-level netlist for your design (ASIC or SOC), then you can start the physical design process.

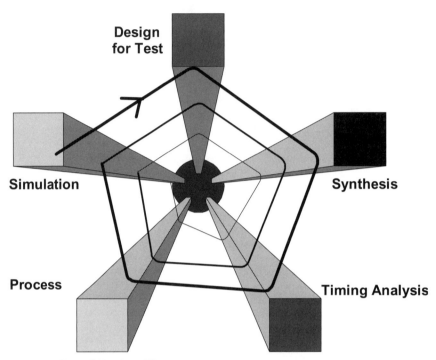

Fig. 1.8 Spiral Design Flow

Figure 1.9 shows a generic physical design, or back-end flow. The major steps consist of place and route, timing verification, and physical verification.

The inputs to place and route are netlist, clock definition, and I/O specification. The goal of place and route is to generate a GDSII file for tapeout. The place-and-route step performs placement, routing, clock-tree synthesis, optimization, and delay calculation.

Task automation is covered in Chapter 3.

1.6 SUMMARY

In this introductory chapter, we defined an SOC and some of its differences from a traditional ASIC. A key concept in SOC design is

1.6 Summary

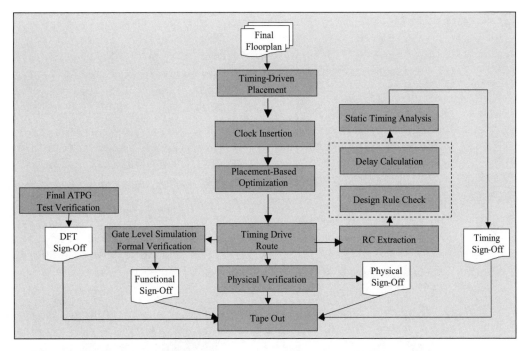

Fig. 1.9 Generic Physical Design Flow (Printed with permission of Fujitsu Microelectronics America, Inc.)

the usage of different IPs. This by itself creates a big challenge in SOC design, namely IP integration.

Reuse methodology is an important factor in SOC designs that reduces time-to-market (TTM). We cover more on ASICs and SOCs, including verification techniques, in Chapters 2 and 3, respectively.

Chapter 4 deals with the physical design domain that is common to both ASICs and SOCs. Once you have a netlist for the proposed IC (ASIC or SOC), then you enter the world of the physical domain.

Chapter 5 covers low-power design concepts and techniques that again are common to both ASICs and SOCs. Several methods of power optimization at different levels of abstraction will be covered. These techniques include algorithm, architecture, Register Transfer, and gate-level optimizations.

1.7 References

1. M. Keating and P. Bricaud. *Reuse Methodology Manual for System-on-a-Chip Designs.* Norwell, MA: Kluwer Academic Publishers, 1998.
2. F. Nekoogar. *Timing Verification of Application-Specific Integrated Circuits (ASICs).* Upper Saddle River, NJ: Prentice Hall PTR, 1999.
3. P. Rashinkar, P. Paterson, and L. Singh. *System-on-a-Chip Verification Methodology and Techniques.* Norwell, MA: Kluwer Academic Publishers, 2001.
4. H. Chang, L. Cooke, M. Hunt, G. Martin, A. McNelly, and L. Todd. *Surviving the SOC Revolution: A Guide to Platform-Based Designs.* Norwell, MA: Kluwer Academic Publishers, July 1999.
5. S. Azimi. *Overcoming Challenges and Obstacles to System on Chip (SOC) Products.* Sunnyvale, CA: Marvell Semiconductor, Inc., 2000.
6. A. Qureshi (Cadence Design Systems, Inc.). "SOC Design Methodology and Ideal Structures." DesignCon2000.
7. D. Wingard. "Integrating Semiconductor IP Using microNetworks, ASIC Design." Mountain View, CA: Sonics, Inc., 2001.
8. R. Fehr. "Intellectual Property: A Solution for System Design." Technology Leadership Day, October 2000.
9. P. Levin and R. Ludwig. "Crossroads for Mixed-Signal Chips." *IEEE Spectrum*, March 2002.
10. R. Rajsuman, *System-on-a-Chip Design and Test.* Santa Clara, CA: Artech House Publishers, 2000.

CHAPTER 2

Overview of ASICs

2.1 INTRODUCTION

ASICs are logic chips designed by end customers to perform a specific function for a desired application. ASIC vendors supply libraries for each technology they provide. In most cases, these libraries contain predesigned and preverified logic circuits.

Several ASIC technologies exist. These are mainly gate array, standard cell, and full custom. Some features of these ASIC technologies are summarized in Table 2.1.

See reference 3 for more details on these technologies. Typical features of ASIC devices with specific technologies that are provided by ASIC vendors are:

- ☞ AC characteristics—AC characteristics, or propagation delay time (t_{pd}), are specified for minimum, typical, and maximum values. These values are determined by wiring capacitance and resistance. Also, junction temperature, power supply voltage, and process variations are used to calculate AC characteristics.

- ☞ DC (static) characteristics—This data specifies the minimum, typical, and maximum values for high-level and low-level out-

Table 2.1 ASIC Technologies

Feature	ASIC Technology Type		
	Gate Array	Standard Cell	Full Custom
Complexity	Medium	Medium to Very High	High to Very High
Speed	Moderate to Fast	Fast	Fast
Development Cost	Moderate	High	High
Available I/O	High	High to Very High	High to Very High
Custom Mask Layers	Some	All	All

put voltage and high-level output current as well as output short-circuit current and input leakage current. These values collectively assure the worst-case values of the DC characteristics of input and output buffers at the operating conditions.

- Recommended operating conditions—These usually consist of minimum, typical, and maximum values for supply voltage, high-level input voltage, low-level input voltage, and the junction temperature. These values are recommended for normal operation of the device.

- Power consumption—ASIC vendors provide formulas to calculate chip power consumption. This is usually determined by the sum of power consumption of I/O buffers, internal logic gates, and on-chip memory. Tools are also available to predict power consumption of ASICs.

- Available packages—Examples are through-hole (dual in-line, or DIP, in both plastic and ceramic; pin grid array, or PGA, in both plastic and ceramic) and surface-mount (quad flat pack, or QFP; small outline package, or SOP; and ball grid array, or BGA). There are advantages and disadvantages with each package and the chip designer should carefully consider what package to select for the specific design.

- ☞ Available macros—Macros are available, ranging from basic logic gates (e.g., AND, OR, NAND, NOR, XOR), latches and flip flops, buffers, adders, multiplexers, synchronous and asynchronous memories, to more complex cores such as CPUs, DSPs, and memory controllers.

- ☞ Types of I/O buffers—Selection of the appropriate input and output buffers depends on interface level, logic function, interface function, pull-up/pull-down option, and drive capability. Examples of I/O buffers are input buffers, input buffer inverting, bidirectional output buffers, and 3-state output buffers.

- ☞ Power on/off sequence—The sequence specifies the correct and recommended power on/off sequence for dual power supply devices as well as for internal and external power sources. Restrictions on external signal levels are also provided by ASIC vendors.

- ☞ Analog cells—Typical analog cells used in an ASIC device include OPAMPs, digital-analog converters (DAC), analog-digital converters (ADC), and phase locked loops (PLL).

- ☞ PLLs—PLLs are used for reduction of on-chip latency, synchronization of clocks between different ASICs, frequency synthesis, and clock-frequency multiplication. Refer to Appendix C for more information on PLLs.

- ☞ Pin assignment rules—Assignment rules for clock, clear, preset input, and simultaneous switching output pins as well as for power and ground pins are also specified by ASIC vendors.

Other technology-related information provided by the ASIC vendors for a specific ASIC technology includes the number of metal layers, the power supply for the core and the I/Os, the junction temperature, and the electrostatic discharge (ESD) specification.

In Chapter 1, we mentioned some of the factors that affect the TAT. The time it takes semiconductor vendors to make an ASIC prototype and a working part is usually referred to as the TAT, or more precisely TAT is the time taken from gate-level netlist to metal mask-ready stage. Figure 2.1 shows the degree of difficulty for TAT. The factors involved for an ASIC TAT include the following:

- Frequency of operation
- Number of gates
- Density
- Number of clock domains
- Number of blocks and sub-blocks

Each one of these factors directly affects the TAT. The higher the factor, the longer the TAT. The customer-vendor relationship and a clear line of responsibility also affect the TAT.

Section 2.2 covers the methodology and front-end design flow for the ASICs. Some useful guidelines are presented for the ASIC

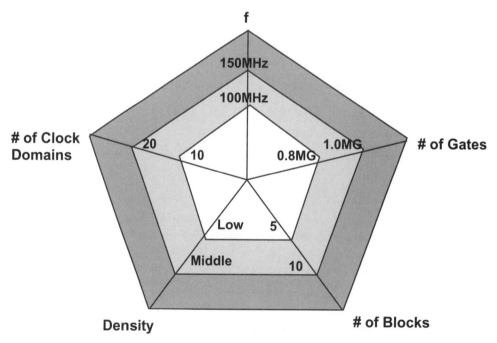

Fig. 2.1 Degree of Difficulty (Printed with permission of Fujitsu Microelectronics America, Inc.)

methodology. Here we assume the designers use Synopsys Prime-Time as the chip-design industry standard STA tool. Some key questions that ASIC designers must consider early on when planning for design are also covered.

In this chapter, we don't cover FPGAs. However, in Section 2.3 we discuss FPGA to ASIC conversion, which is becoming more popular among system designers for cost-cutting purposes.

An overview of the verification methodologies is covered in Section 2.4.

2.2 METHODOLOGY AND DESIGN FLOW

As mentioned in Chapter 1, a good design methodology consists of a set of defined design flows for both front and back end as well as for tool integration and task automation. Figure 1.7 in Chapter 1 showed the basic front-end ASIC design flow. This figure is repeated as Figure 2.2.

Major timing issues should be resolved here at the front end, before performing a detailed floorplan. In Chapter 4, we cover back-end or physical design flow, where you resolve minor timing issues.

Here at the front end, we start with RTL coding. This can be done in Hardware Description Languages (HDL) such as Verilog or VHDL. Lint tools should be used to check the RTL code for any coding and syntax violations. Functional simulation is the first necessary step after RTL coding is completed. Functional simulation verifies the design for its functional requirements as they are defined in the specs.

Synthesis translates a specific abstraction into the next level. For example, behavioral synthesis translates behavioral HDL to RTL architecture. Logic synthesis translates RTL into a technology-specific design in the form of gate level. Major steps in logic synthesis are translation, mapping, and optimization. The user can set his or her own constraints on area, speed, power, routability, and testability. Appropriate wire-load models (WLM), aggressive or conservative, have to be specified for the target design.

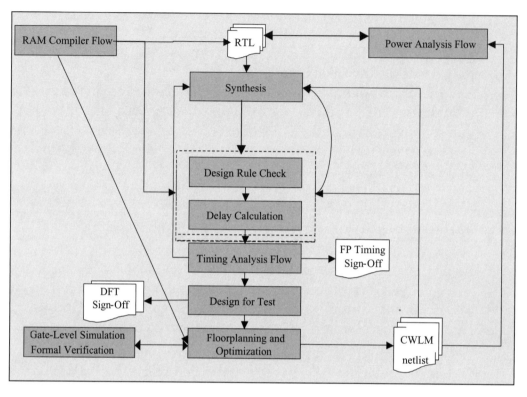

Fig. 2.2 Front-End Design Flow (Printed with permission of Fujitsu Microelectronics America, Inc.)

Design rule constraints must take precedence over optimization constraints because if the gates of a library cannot meet the designer's requirements, there is nothing that can be done except to get a higher performance library. The following three constraints, or specifically Design Compiler commands, should be used to ensure that the limits of the ASIC library are not exceeded:

- set_max_fanout
- set_max_transition
- set_max_capacitance

Power analysis is an important part of the design process. Power-estimation as well as power-reduction techniques are done at different levels of the design abstraction, such as at RT and gate levels. We cover power analysis in Chapter 5.

The first part of timing analysis is RC extraction, which extracts parasitic information from interconnect wires. The second part is static timing analysis (STA), in which delay is calculated and timing constraints of the chip are verified. STA verifies the delays within the design. It is capable of verifying every path and can detect serious problems like glitches on the clock, violated setup and hold times, slow paths, and excessive clock skew.

DFT techniques such as scan, Automatic Test Pattern Generation (ATPG), and Built-In Self Test (BIST) are applied to the ASIC. We discuss DFT in greater detail at the end of this chapter.

Floorplanning is discussed in Chapter 4.

Some Useful Guidelines

For designs bigger than two million gates in size:

- Use hierarchical methodology. Define sub-block boundaries and sub-block pin assignments. Use straightforward clock structure. Although clock-gating is an attractive low-power design technique, it can cause low clock skew for clock tree.

- Use placement-based synthesis. This improves timing and congestion problems in the layout phase. Placement-based synthesis is applied to the pretest and preclock design. Synopsys Physical Compiler and Cadence Envisia PKS are examples of placement-based synthesis tools. Figure 2.3 shows a typical design flow with placement-based synthesis.

- When applying hierarchical methodology, use Interface Logic Models (ILM) for Primetime timing analysis. These models improve the performance of chip-level timing analysis mainly by reducing the size of the netlist. ILMs have replaced the traditional STAMP models. Figure 2.4 shows how ILMs are used in chip-level verification. In this figure, the internal register-to-register logic is discarded for the blocks. However, I/O logic and clock tree are preserved.

28   Overview of ASICs   Chap. 2

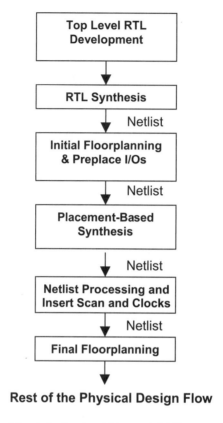

Fig. 2.3 Design Flow with Placement-Based Synthesis

☞ In hierarchical design methodology, three different types of timing constraints have to be defined by the logic designer (not by the physical designer). These are:

1. Chip-level timing constraints for the final timing verification.
2. Chip-level timing constraints excluding the hierarchical logic blocks (HLB). These are blocks that can be laid out independently as hard macros.

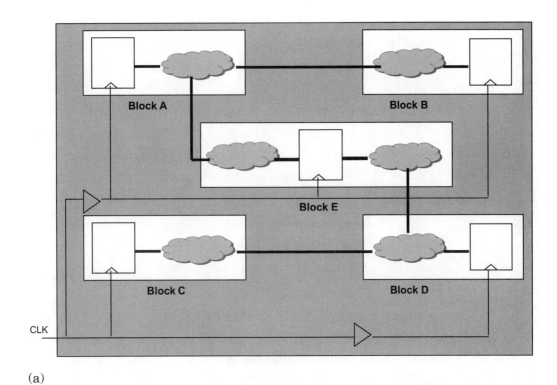

Fig. 2.4 (a) ILM models in Chip-Level Timing Verification (b) Inside of an ILM Model

3. Timing constraints for the HLBs.

Figure 2.5 shows the three types of the timing constraints.

☞ Execute static timing analysis at chip-level for the following conditions:

1. System mode (min and max)
2. Test mode (min and max)

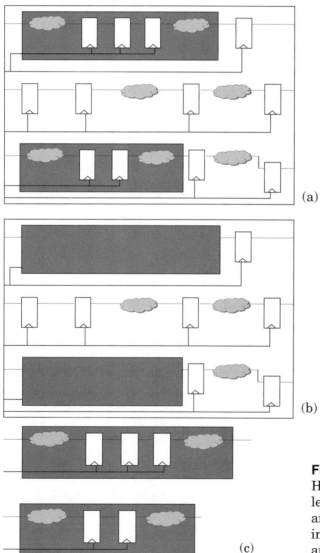

Fig. 2.5 Timing Constraints in a Hierarchical Design (a) Chip-level analysis, (b) Chip-level analysis excluding the timing inside HLB, and (c) Timing analysis inside the HLB

Key Questions for Your Design

When designing ASICs, there are a lot of issues and factors that need to be addressed early in the process. These are in addition to other well-known factors with RTL coding, logic design, timing analysis, design flow, tools, verification, and physical design issues which are covered in Chapter 4. An ASIC team should consider the following key questions:

- What kind of system or end application is the ASIC designed for?
- Does the chip have previous design? In the case of porting, what is the methodology for the design database migration including netlist and vector translation?
- What is the power supply voltage? What is the tolerance level?
- How much memory and what kinds are required? For example, what is the bit/word configuration for SRAMs? How many instances per chip? What are the speed, access time, and cycle time? How many read/write ports?
- What will be the power consumption? Have a clear idea on the power consumption and power requirements of the ASIC you are designing; the amount of power the chip consumes affects the board and the system-level designs.
- What is the die size estimation? The designer must perform die size estimation early on. The size of the die directly affects the power, timing, and routability and hence cost, effort, and the design schedule.
- What is the package type? Can the package handle the required power?
- How many high-speed signal pins does the design have?
- What kind of interfaces are you designing for your ASIC? Is it asynchronous or source (clock) synchronous?

☞ What kind of standards are you required to follow for the high-speed interface macros of your design? Examples of these standards include the following:

SFI-4 and SFI-5 from Optical Internetworking Forum
10G Ethernet for LAN, MAN, and WAN applications
IEEE802.3ae (an IEEE LAN standard)
SPI-4P2 from Network Processing Forum

2.3 FPGA TO ASIC CONVERSION

In order to bring down the cost of systems, vendors convert their FPGA designs into ASICs. In some cases, multiple FPGAs are integrated into a single ASIC. Some ASIC vendors match the exact package and pinout of the original FPGA designs.

The market targeted for these migrations is the middle ground of the ASIC market. ASIC variations that are offered include gate array/embedded array and standard-cell technologies, multimetal-layer processing, high-speed operation, low-power and low-voltage operation.

Figure 2.6 shows a typical conversion flow from FPGA to ASIC. The FPGA netlist typically comes in the form of Verilog, EDIF, VHDL, or XNF.

Design analysis usually consists of boundary scan, power analysis, design-rule checks, pin-pad selection, and package finalization.

Depending on the complexity of the design and its test vectors, the conversion flow changes slightly. ASIC design houses usually have defined flows for both synchronous and asynchronous designs with full or partial test vectors.

In addition to cost reduction, other benefits of the conversion include the following:

☞ Die size reduction
☞ Power consumption reduction
☞ Enhanced performance

2.3 FPGA to ASIC Conversion

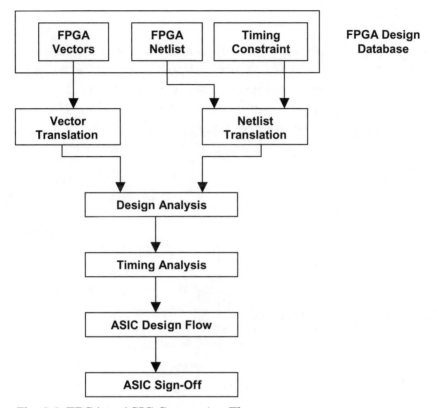

Fig. 2.6 FPGA to ASIC Conversion Flow

- ☞ Reliable high-volume production capacity
- ☞ Low NRE

The following tips should be considered for the FPGA to ASIC migration:

- ☞ Some features, such as RAM initialization and configuration logic, are expensive to implement in an ASIC. These features should be avoided in ASICs.
- ☞ Use synchronous design methodology.
- ☞ Leave extra power and ground pins for your pinout.
- ☞ Use one external clock and external reset.
- ☞ Use standards as much as possible.

2.4 VERIFICATION

Today about 70 percent of design cost and effort is spent on verification. Verification teams are often almost twice as large as the RTL designers at companies developing ICs. Traditionally, chip design verification focuses on simulation and regression. However, new verification techniques are emerging. Figure 2.7 shows several

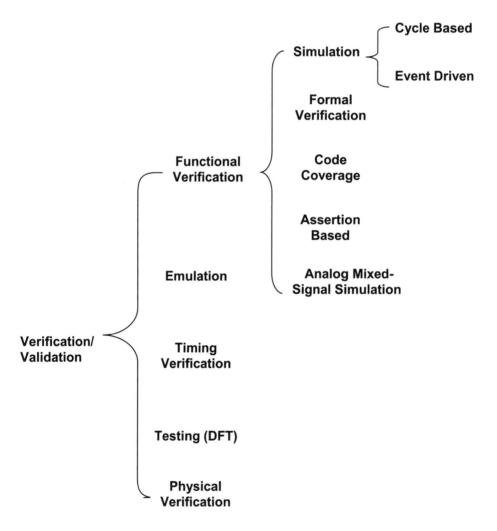

Fig. 2.7 Types of Design Validation and Verification

types of design verification and validation, which are also described here.

Functional Verification

Functional verification is an important part of design validation. Functional verification must be done at all levels of hierarchy in the design. It must be done at both chip and block levels.

Simulation

Simulation techniques are used at both RT and gate levels. Depending on the size of the design entity and the simulation run time, the RTL designers choose between cycle-based (where all nodes are evaluated at the cycle boundaries) and/or event-driven (where all nodes are evaluated at every event) simulators.

Testbenches

Testbenches can be categorized into the following three major categories:

- ☞ Vector-based. This is the traditional way of writing testbenches where the stimulus and the response are in the form of vectors.
- ☞ BFM-based. The bus functional model-based test benches are easier to write sequences for and can be reused. However, it is hard to predict the device response.
- ☞ C-based. This is a high-performance algorithm execution that uses C-language constructs. However, C is not designed for hardware verification.

Testbench creation and automation tools are becoming very popular among design and verification engineers. Two examples are Vera from Synopsys and Specman Elite from Verisity.

Formal Verification

Formal verification is a mathematical method for determining the equivalence of two representations of a circuit. It is mainly used for RTL-to-RTL, RTL-to-gate, and gate-to-gate comparisons. Formal verification doesn't use any test vectors and is much faster than conventional simulation techniques. This is an effective method of verification after insertion of test logic and clock tree into the original circuit. EDA tools from Synopsys and Verplex, Inc., are examples of formal verification tools.

Code Coverage

An effective way of checking RTL is code coverage. Code-coverage tools essentially check the RTL code to find out what portions have been exercised in the design. Types of coverage tests include state machine, branch, path, expression, condition, triggering, and toggle.

Assertion-Based Verification (ABV)

This traditional black-box approach to verification discovers bugs at the boundaries of the box. ABV uses a so-called white-box approach to verification. Here the user can discover internal bugs early on, before they propagate outside of the box and corrupt the rest of the system. Assertion-based verification is used at RT level for both Verilog and VHDL. ABV tools are available from Cadence Design Systems and 0-In Design Automation, Inc.

Analog Mixed-Signal (AMS) Simulation

AMS simulation deals with both analog- and digital-simulation techniques to verify a circuit. Therefore, it is more complex than either digital or analog techniques alone. The analog blocks are usually modeled and verified using predefined models of basic elements (resistor, capacitor, etc.) with SPICE simulators. Some common

SPICE simulators are H-SPICE, P-SPICE, Z-SPICE. Although SPICE simulators are very accurate, they are not fast simulators. Once an analog block is verified, its interface will be dealt with as a digital interface in an SOC-integration process.

Emulation

In-circuit emulation is the process of running a software model of an ASIC with the real system. Emulation is much faster than simulation and is an attractive method for large ASICs, providing the entire design is placed in the emulation engine. However, emulation is not used for block-level verification, is expensive, and is used late in the design-verification process.

This method of design validation is not suitable for detailed design debugging.

Design for Test (DFT)

DFT is an important part of the design and validation process. DFT techniques were traditionally performed before the floorplanning step (see Figure 2.2). However, as part of the new design methodologies, DFT techniques should be considered and applied at several points in the design process. For example, scan insertion can be integrated into the synthesis and physical placement or ATPG is sometimes integrated with the routing process.

DFT is a vast topic, and there are whole books dedicated to the subject (for one example, see reference 8). Here, we offer a basic overview of DFT. Testing methods usually consist of boundary scan, ATPG, and memory testing techniques such as BIST.

The goal of boundary scan is to make sure that every node in the design is both controllable (controlling a node to a specific state from the primary inputs) and observable (observing the state of a node at the primary outputs) through testing methods.

The IEEE 1149.1 standard defines boundary scan as one with Test Access Port (TAP). This standard is also known as the JTAG standard, named for the Joint Test Action Group that initiated these standards on testing. The basic technique of boundary scan can be applied to both chips and boards. The dedicated test pins for TAP are as follows:

- TRST (Test Rest)
- TMS (Test Mode Select)
- TDI (Test Data In)
- TDO (Test Data Out)
- TCK (Test Clock)

Using TAP port you can access the internal logic and pins of a chip. Figure 2.8 illustrates a generic device under test with scan cells inserted for every I/O pad in the device. Each scan cell is a storage element capable of data multiplexing. The cells are connected together to form a shift register that goes around the chip.

Test vectors generated by ATPG tools are applied at the scan cell inputs and the response is captured at the scan cell outputs.

Aside from being able to directly observe and control internal nodes, boundary scan is an efficient as well as an automatic testing method. However, there are few drawbacks with scan testing. These drawbacks include the following:

- Additional time and effort are required, affecting the TAT.
- Additional gates are required, causing an increase in the final die size. In some designs you may end up with almost 20 percent of your total logic gates dedicated to DFT logic.
- Additional dedicated I/Os are required. This impacts package pins.
- Elevation of the overall chip power consumption is required.
- Negatively impacts timing.

ATPG tools generate test vectors using algorithmic methods. There are techniques available to convert RTL simulation results into files that ATPG tools can use to generate scan patterns (see ref-

2.4 Verification

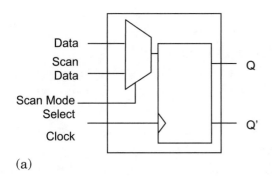

(a)

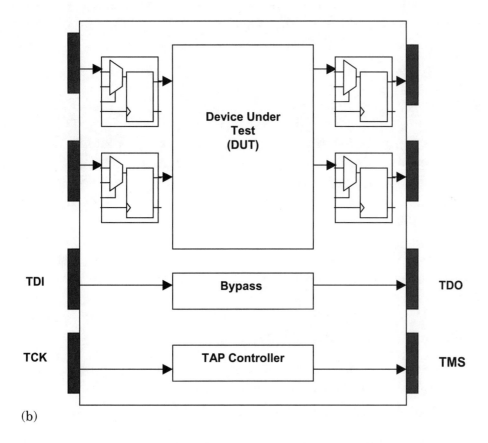

(b)

Fig. 2.8 Device Under Test with Inserted Boundary Scan Cells: (a) Scan cell (b) Boundary scan

erence 7 for more on this subject). ATPG tools differ from one to another depending on the algorithm they execute. Examples of these algorithms are combinational and sequential.

BIST is an effective DFT method for logic and memory elements such as RAMs and ROMs. BIST reduces the need for external testers. An ASIC can test itself once certain logic (BIST logic) is added to the chip. Similar to scan insertion, implementing BIST requires additional logic overhead.

When you meet all your DFT requirements, then the design is ready for DFT sign-off. DFT tools from LogicVision, Inc., and Mentor Graphics, Corp., are the most popular ones among chip designers.

2.5 SUMMARY

Table 2.1 compared complexity, speed, cost, and available I/Os for gate arrays and standard-cell and full-custom ASICs.

The major factors involved in a design TAT are frequency of operation, number of gates, density, number of clock domains, and number of blocks and sub-blocks.

The basic front-end design flow was discussed in Section 2.2. Hierarchical methodology, the use of placement-based synthesis, and the use of ILM models are recommended for designs bigger than two million gates.

FPGA to ASIC conversion was discussed in Section 2.3. Some benefits of the conversion include:

- Die size reduction
- Power consumption reduction
- Enhanced performance
- Reliable high-volume production capacity
- Low NRE

An overview of the verification techniques and DFT was discussed in Section 2.4. We will cover more on verification issues in Chapter 3.

2.6 REFERENCES

1. J. Bergeron. *Writing Testbenches*. Norwell, MA: Kluwer Academic Publishers, 2000.
2. *www.cis.ohiostate.edu / ~harrold / research / regression_testing.htm*
3. M. J. S. Smith. *Application-Specific Integrated Circuits*. Reading, MA: Addison-Wesley, 1997.
4. D. Hsu. "DFT Closure in SOC Design." Synopsys, Inc., Mountain View, CA, 2000.
5. J. Desposito. "SOC and Deep-Submicron Technology Drive New DFT Strategies." Electronic Design, 1998.
6. "JTAG Boundary Scan Testing." Testability Primer, Texas Instruments, 1997.
7. B. Murray and J. Hayes. "Testing IC's: Getting to the Core of the Problem," *Computer*, Vol. 29, No. 11, IEEE, November 1996.
8. A. Crouch. *Design for Test*. Upper Saddle River, NJ: Prentice Hall, 1999.
9. "Solving the Challenges of Testing Small Embedded Cores and Memories Using FastScan Macro Test." Mentor Graphics, 2000.
10. M. Keating and P. Bricaud. *Reuse Methodology Manual for System-on-a-Chip Designs*. Norwell, MA: Kluwer Academic Publishers, 1999.
11. F. Nekoogar. *Timing Verification of Application Specific Circuits (ASICs)*. Upper Saddle River, NJ: Prentice Hall PTR, 1999.
12. Spring Tech Seminars, Verification 2001, Synopsys, Inc., Mountain View, CA.
13. ASIC Products Application Notes. "Application of Synopsys Physical Compiler in IBM ASIC Methodology." IBM, August 2001.
14. ASIC Products Application Notes. "Application of Cadence Envisia PKS in IBM ASIC Methodology." IBM, May 2001.
15. CMOS ASIC CS81/CE81 Series, Product information notes. Fujitsu, March 2000.

16. "FPGA to ASIC Conversion Program." Epson Electronics America, Inc., November 2000.
17. *www.amis.com/conversion*
18. "Hierarchical Static Timing Analysis Using Interface Logic Models, PrimeTime." Synopsys, Inc., Mountain View, CA, January 2001.
19. "Best Practices and Advanced Verification Techniques." Spring Tech Seminars, Verification 2001, Synopsys, Inc., Mountain View, CA.
20. "Complete Best-in-Class Synthesis Solution." Spring Tech Seminars, Synthesis 2001, Synopsis, Inc., Mountain View, CA.
21. *www.cadence.com/datasheets/assertion_based_verification.html*
22. *www.cadence.com/products/fv.html*
23. N. Deo. "Tools and Technologies for Deep Submicron IC Design." 1997.

CHAPTER 3

SOC Design and Verification

3.1 Introduction

In Chapter 1, we introduced the concept of SOCs and IPs. We also discussed some of the SOC design challenges. In this chapter, we complete our discussion of SOCs. We will cover physical design issues related to SOCs in Chapter 4.

Section 3.2 covers design for integration. Here we cover more about on-chip buses (OCBs) and continue with the example we used in Chapter 1 on VoIP.

Section 3.3 discusses SOC verification. The verification methods that we covered in Chapter 2 for ASIC design verification can also be used for SOCs. However, we will discuss additional issues affecting SOCs, in particular those that are due to the usage of several IPs on an SOC. The earlier these issues are addressed in the SOC design cycle, the faster and more complete will be the verification process.

We end this chapter with a complete example for an SOC. Section 3.4 covers a set-top box (STB) design example. The example shows architectural investigation and modeling to show the efficacy of a specific on-chip communication network as a unifying communications medium and it shows a new, unified memory architecture with end-to-end performance guarantees.

3.2 DESIGN FOR INTEGRATION

As we mentioned in Section 1.3, a key issue in SOC design is integration of silicon IPs (cores). Integration of IPs directly affects the complexity of SOC designs and also influences verification of the SOC. Verification becomes faster and easier if the SOC interconnect is simple and unified, as was the solution introduced in Chapter 1 for system integration (i.e., use an on-chip communication system or intelligent on-chip bus).

In general, there is no standard for OCBs; they are chosen almost exclusively by the specific application for which they will be used and by the designer's preference. Two main types of OCBs and their characteristics are shown in Table 3.1.

Table 3.1 OCBs and Their Characteristics

OCB	Speed	Bandwidth	Arbitration	Example
System	High	High	Complex	ARM AHB
Peripheral	Low	Low	Simple	PCI Bus

A detailed discussion on design of buses and their architectures is beyond the scope of this book. Here, we are mainly interested in using an OCB for system integration.

A typical SOC uses both system and peripheral OCBs. There are also a lot of other wires, such as dedicated links and control signals, between blocks. This is shown in chapter 1, Figure 1.5.

More on VoIP SOC

Figure 1.1 showed the block diagram of a gateway VoIP SOC. We also defined its subsystems. Let's examine this SOC more carefully and see how we can use an intelligent OCB for this VoIP SOC. A

3.2 DESIGN FOR INTEGRATION

gateway VoIP SOC is a device that is used for carrier VoIP gateway functions such as vocoders, echo cancellation, data/fax modems, and VoIP protocols. Currently, there are a number of these devices available from several vendors and, typically, these devices differ from each other depending on the type of functions and voice processing algorithms they support.

Bridging of traditional circuit-switched voice to packet-switched networks is handled by mediation gateways. These are usually proprietary box designs with specialized backplanes. These systems include a chassis and a variety of optional interfaces that can be configured based on customer requirements. Voice processing takes place on a voice-processing subsystem, which is a PCB, attached to the system backplane.

In order to reduce the size of these subsystems, SOCs are used in such systems to provide voice-processing and packet-processing functions to bridge between traditional TDM data and the emerging IP/ATM transport backbone.

Table 3.2 shows some of the key features and their requirements for a mediation gateway SOC voice processor.

Figure 3.1 shows a VoIP SOC architecture using SiliconBackplane. Here, we have multiple-processor cores (CPU is used for

Table 3.2 Mediation Gateway SOC Voice Processor Feature and Requirement Summary

Feature	Requirement
CODECS	G.711, G.729, G.726
Transport/Packet	AAL1, AAL2, UDP, TCP, IP, TDM
Stream I/O	H.100/H.110, TSI, Utopia, Ethernet
Host I/O	PCI/Synch Proprietary
Other	Memory: 8–16MB, SDRAM, 32 BIT Flash:4 – 8 MB Flash/ROM
SOC Interconnect	Proprietary
Tone Processing	DTMF, MF
Line Echo Canceller (LEC)	G.168 Compliant

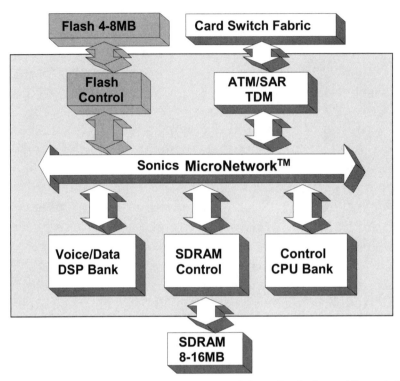

Fig. 3.1 VoIP SOC Architecture Using SiliconBackplane (Copyright 2002, Sonics, Inc.)

packet protocols and overall control function), multi-DSP cores (DSP is used for voice and modem protocols and LEC), SDRAM (serves all the cores), and I/Os (flash control for card boot up and TDM controller) all unified by SiliconBackplane.

A typical voice-to-packet flow consists of the following steps:

1. Voice port receives voice or fax.
2. Voice port writes frames to memory.
3. Each packet processor reads data, processes data, and then writes data back into memory for the next processor.
4. Packet port reads packets from memory.
5. Packet port sends packets out onto packet network.
6. CPU controls all the above transfers.

And a packet-to-voice flow includes the following steps:

1. Packet port receives packets from packet network.
2. Packet port writes packets to memory.
3. Each packet processor reads data, processes data, and then writes data back into memory for the next processor.
4. Voice port reads frames from memory.
5. Voice port sends voice or fax out onto voice/fax interface.
6. CPU controls all the above transfers.

In this design we can go up to 4GB/s SiliconBackplane BW (128bits, 250MHz).

Some of the benefits of using SiliconBackplane in a VoIP SOC are as follows:

- Fast and early architectural prototyping. This decreases design-cycle times, promotes early optimization, increases confidence, and improves verification time.
- High reusability of cores. This will improve your TTM.
- Late-stage architectural trade-offs and ECOs. You can insert additional functional units into the SOC architecture plug-and-play. For example, you can reduce or increase processing power by inserting a new CPU into the architecture.
- High performance. You can increase the number of channels to 1000 and beyond.
- Flexibility. You can scale down to small, low-cost, point solutions.

3.3 SOC Verification

All the verification methods we covered in Chapter 2 for ASICs apply to SOCs as well. However, SOC verification becomes more complex because of the many different kinds of IPs on the chip. A verification plan must cover the verification of the individual cores as well as that of the overall SOC. A good understanding of the overall application of the SOC is essential. The more extensive and intense the knowledge of the external interfaces and their interactions with the SOC, the more complete the SOC verification will be. Various SOC applications require unique external interface constraints, and the verification team should consider those constraints early on.

The SOC functionality can be represented with data-flow and control-flow models.

Data-flow analysis determines the bandwidth capacity of an SOC interconnect and the requirements of its various components by considering the amount of data that needs to be processed under real-time conditions. This is because in many situations sources of data and the final processed data could have different rates.

Control-flow analysis for an SOC takes into account the nature and rate of external interface processing. The control of the data and events from outside may be in various time domains, or it may be totally asynchronous in nature. The rate of change on these interfaces may also need to be determined to allow for sufficient processing power within the SOC to be able to respond within given timing and data-flow constraints. Some responses may need software processing and/or other external interface dependencies.

A detailed data-flow and control-flow analysis for a given SOC will result in the necessary verification strategy.

Verification Planning Guidelines

The following should be considered in verification planning:

- **External Interface Emulation** When verifying complex SOCs, in addition to logic simulation techniques full chip emulation should be considered. The primary external interfaces of each IP, as well as the SOC data interfaces, should be examined to evaluate the need for any SOC simulation. This should be performed simultaneously for all cores in order to evaluate the modeling work required early on.

3.3 SOC VERIFICATION

☞ **Hardware/Software (HW/SW) Integration** HW/SW integration must be planned for SOCs with processor type cores. Developing bus-functional models should be part of the plan.

☞ **Verification Resource Planning** Resource planning is a vital step for a successful SOC verification. The size of a verification task can predict the simulation hardware resources and the needed personnel. Similarly, the number and complexity of IPs in an SOC will determine the amount of estimated regression time, hardware computing resources, and simulation license requirements.

Since in a complex SOC there are many IPs and each one may have several versions, hence source code control software such as CVS and RCS must be used in the verification environment. Source code control software keeps track of any changes made to various design files. It also protects the directories and files from accidental deletion. When multiple engineers work on the same file, source code control keeps a record of who did what and when.

For large and complex test files of an SOC with several levels of hierarchy, Makefiles are essential. The `make` utility keeps track of which files need to be updated and recompiled. Thereafter, a simple `make` command performs the updating and recompiling job. This can provide substantial reliability in verification.

A plan has to be developed to identify which of the cores have to be interfaced and accessed through the CPU. Therefore, a list of tests must be defined to represent the actual firmware that will be used in the final SOC system software.

☞ **Regression Planning** Regression testing is the process of verifying designs to guarantee that earlier debugging has not affected the overall functionality. It's quite possible that a change in a design made to fix a problem, detected by a (debugging) test case, may in fact break the functionality that was previously verified. Therefore, regression testing should not be confused with debugging. Another benefit of regression testing is a reassurance that the design is backward compatible with the original design that was tested. Typical regression tests run anywhere from 15 hours to a few days. HDL languages such as Verilog and VHDL are widely used in regression testing. These languages offer a faster approach that can speed up the regression testing period. The user command line option in Verilog for this approach is `fastmod`.

It should be noted that regression testing can run indefinitely if the condition it is waiting for never happens. Therefore, a mechanism to prevent such endless runs is to automatically terminate the testing after a specific amount of time. This can be done in VHDL using the time bomb procedure.

Regression testing can be automated by using batch files and scripts to provide more reliability for complex SOCs. In automated regression, various tests can be distributed over multiple workstations and the differences between the report files can be stressed.

In addition to the above verification planning guidelines, the verification team should also consider both dynamic timing simulation and static timing verification.

Dynamic timing simulation should be used for timing analysis of asynchronous designs as well as synchronous designs. Using dynamic simulation the verification team can verify the functionality as well as the timing requirements of a design. The team must develop comprehensive input vectors to check the timing characteristics of critical paths in a design.

Static timing analysis should be used to verify the delays within the design. Using STA, the design team must verify every path and detect serious problems such as glitches on the clock, violated setup and hold times, slow paths, and excessive clock skew.

Figure 3.2 summarizes the verification planning elements.

3.3 SOC Verification

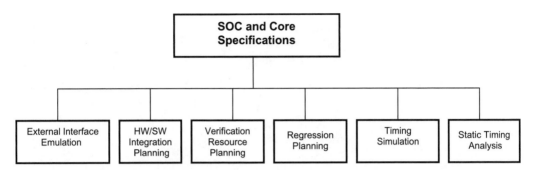

Fig. 3.2 SOC Verification Planning

SOC Verification Execution

What we have covered so far in this section dealt with preparation and planning for verification. Here, we outline some tips for SOC verification execution.

- ☞ The verification team should pay special attention to the power-up and power-down sequencing of the different cores in the chip, both during simulation and during device bring-up.
- ☞ The register inside each core should be carefully verified.
- ☞ Individual cores should be tested. Regression, debugging, and test coverage should be performed on all individual cores.
- ☞ 100 percent code coverage is desirable. Low code-coverage numbers should alert the verification team that additional testing is required.
- ☞ Software reuse can speed up the verification process in device bring-up.

Figure 3.3 illustrates the SOC verification execution flow.

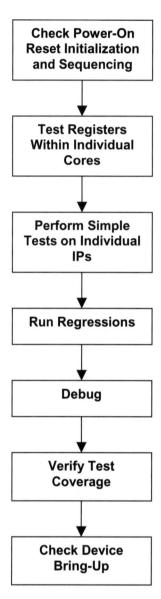

Fig. 3.3 SOC Verification Execution Flow

IP Verification

Reusing previously verified IPs can tremendously expedite SOC verification. Since IPs can be delivered in different formats from various IP vendors, the verification planning phase should start early on. An IP acceptance checklist should be developed and followed, whenever third-party IPs are used in an SOC design. A typical checklist consists of the following:

1. IP design files
   a. Documentation for the directory structure
   b. Verilog/VHDL RTL source files
   c. Source files for sub-blocks
2. Documentation on revisions
   a. Functions no longer supported
   b. Information on revision control
3. Verification/simulation
   a. Testbench or test cases for regression test
   b. Readme files for running the tests
   c. Information on testbench generation tools, such as Vera and Specman, if used for the IPs
   d. Source codes for any C, C++
   e. Scripts such as Perl
4. Synthesis/technology-related and test coverage
   a. Synthesis scripts for core resynthesis
   b. Functional specifications for the cores/IPs
   c. Specifications on timing, power, and area
   d. Operating conditions and clock speeds
   e. Set-up requirements for all I/O signals
   f. Detailed information and specifications on the sub-blocks
   g. Any information on DFT coverage for the cores

Figure 3.4 shows a simplified overview of IP verification flow. First an IP is verified by the IP vendor, then the user reverifies the IP in his own environment. This is usually a tedious and time-consuming task, because design files and testbenches have to be translated and reverified. Also, in order for the acquired IP to communicate with other IPs on the SOC, the core has to be specially wrapped according to specific interface protocols (see appendix B on Open Core Protocol for one such example). Wrapping an IP core is another tedious task that deals with adding bridges and extra logic to the existing IP. Finally, after the core is successfully wrapped, it is now ready to be integrated with the rest of the SOC.

Automation

Automation is another important part of the overall verification strategy. Automation can help speed up the verification process tremendously. Automation tools usually consist of the following:

- Scripting languages
- Version control systems
- Makefiles

Verification engineers automating repetitive commands into simple scripts use scripting languages such as Perl and Python everyday. UNIX shell scripts are also popular among verification engineers. Examples of UNIX shell commands are Bourne, Korn, and C shells.

As was mentioned, version-control systems should also be used. Examples of such systems include CVS, RCS, and makefiles.

Automation should be applied to every step of the design, verification, and even manufacturing flows. Since EDA tools are constantly evolving, it is important to make sure that communication from one tool to another takes place smoothly, hence not disturbing your flow. Automation techniques are also used to enhance the communication between tools.

3.3 SOC Verification

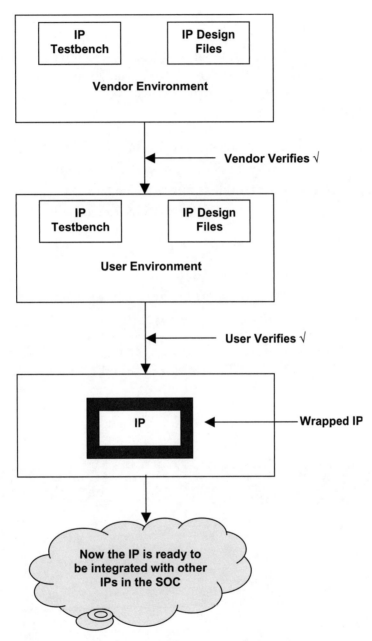

Fig. 3.4 IP Verification Flow

3.4 SET-TOP-BOX SOC

In Chapter 1, we showed an example of a set-top-box SOC. The block diagram for this SOC is repeated here in Figure 3.5.

In this SOC, typical traffic flow for cable applications is as follows:

- From video-processing unit to memory controller
- From memory controller to video-processing unit
- From I/O controller to memory controller
- From memory controller to display controller

For satellite applications, the traffic flow changes to:

- From Utopia interface to video-processing unit
- From video-processing unit to memory controller
- From memory controller to display controller

All of the above traffic goes through the on-chip interconnect. The following example covers a detailed STB SOC with Sonics SiliconBackPlane as the on-chip interconnect.

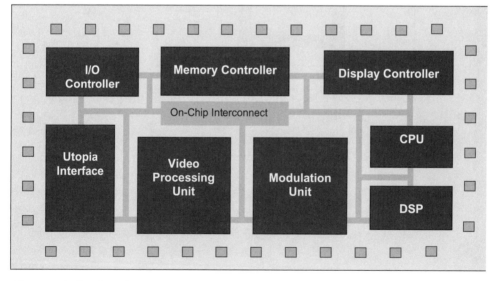

Fig. 3.5 A Set-Top-Box SOC

3.5 SET-TOP-BOX SOC EXAMPLE[1]

Future generations of STBs must embrace SOC technologies in order to deliver new levels of performance while cutting costs.

STB architects and designers face tough challenges in developing increasingly complex systems; these problems are compounded by aggressive TTM goals. STB design challenges can be met only through core reuse, better tools, new design methodologies, and approaches that promote standardization.

During this STB design exercise, a system model was built and simulated using the Sonics FastForward Development Environment. The architecture employs a DRAM subsystem in conjunction with a SiliconBackplane. Among the advantages of this pairing are the following:

- End-to-end performance guarantees. The cores and the SiliconBackplane cooperate to seamlessly provide quality-of-service guarantees for each discrete dataflow.

- Reduced costs through reduced complexity and efficient sharing of memory and interconnect resources.

- High-level functional decoupling enabled by system dataflow service abstractions.

- Predictable timing behavior reduces reliance on overdesign of subsystems. Core data buffers are reduced or eliminated, and buffering requirements overall are more easily assessed.

STB Application Description

This example takes the common digital TV facilities of today's STBs and adds new Internet and multimedia functions. These features include DVD, the Advanced Television Enhancement Forum protocol, time-shifting (instant fast-forward and review, which allows the viewer to be away from the TV without missing any parts of a pro-

---

1. Copyright 2002, Sonics, Inc. Portions reprinted with permission.

gram), VoIP, video telephony, and even home video editing. Family members can be watching TV, Web surfing, recording a favorite program from another TV channel, keeping track of sports events on a third channel, making a VoIP telephone call, and listening to an Internet radio station upstairs over a Bluetooth link all at once. On another occasion, there could be a videophone call taking place while someone reviews a favorite football play using the time shift feature of the STB.

Figure 3.6 provides an overview of the application with the many interface devices and components. Interconnect and unified memory has long been an attractive proposition with its conceptually simple, clean architecture.

STB Characteristics

Delivering multiple functions simultaneously, the STB is a multitasking, software-driven platform with a mixture of general and

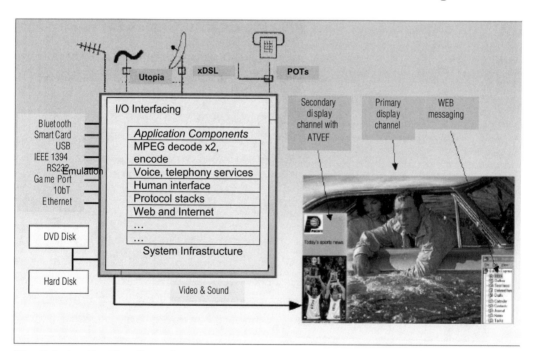

Fig. 3.6 Application Overview

application-specific processors. Numerous processes execute concurrently, most with critical, hard real-time dataflows such as video refresh. To accommodate these processes, both video compression and decompression, plus a host of other digital signal-processing functions, are required.

Though the proposed STB resembles other multimedia platforms, some important factors define and differentiate it. Most importantly, price constraints dictate using low-cost, commodity DRAM in a unified, memory subsystem. STB cost and performance depend on how efficiently the DRAM and communications infrastructure support the mixture of concurrent, high-bandwidth, real-time, and non-real-time traffic.

Dataflows are predominantly between cores and DRAM with little processor-to-processor traffic. DRAM provides the necessary buffering between processes ranging from small bit-stream FIFOs to video frame buffers. To function correctly, interconnect and memory systems must handle the collective peak bandwidth of all real-time flows under worst-case conditions (approaching 1GB/s). The system needs to provide a broad range of performance guarantees with differing quality of service since the demands of individual dataflows vary radically. (Video refresh requires a significant percentage of system bandwidth while audio needs orders of magnitude less.)

The STB shown in Figure 3.7 uses a unified communications backbone, the Sonics SiliconBackplane, to connect all cores. Use of a unified Interconnect and unified memory has long been an attractive proposition with its conceptually simple, clean architecture.

Previous attempts at unification in real-time or multimedia applications have been marred by deficient and unpredictable performance. Design errors in these systems have been recognized as among the most costly to resolve and the hardest to manage. To meet performance goals and retain predictability, typical designs avoid shared resources and employ multiple buses and memories. The new SOC architecture directly addresses many of these issues.

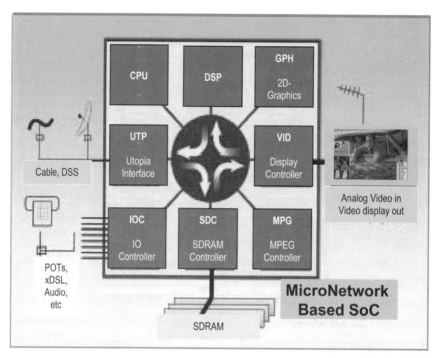

Fig. 3.7 STB SOC System Block Diagram

Dataflow Concepts and Resource Sharing

Resource sharing traditionally trades hardware for behavioral complexity. Increasing the number of sharing or shared units decreases costs, but it increases behavioral complexity by layering extra dimensions of interaction on complex core timings. In this time domain, formalisms, design methodologies, and tools are relatively undeveloped.

A prime objective of this STB project is to reduce complexity by developing a model with dedicated, virtual channels for each dataflow. The SiliconBackplane supports this model with features that enable bandwidth and latency guarantees for each dataflow channeled across it.

The proposed system permits data streams to be analyzed in isolation, processes to execute independently without resource con-

3.5 SET-TOP-BOX SOC EXAMPLE

flicts, and bandwidth to be managed as an assignable commodity. This level of determinism reduces system complexity by recasting the integration challenge as

1. A rigorous definition of dataflows and a determination of their individual requirements
2. Development of a system configuration to serve each and all of those requirements

DRAM Block Accesses

The DRAM controller model used here provides performance guarantees that complement those of the SiliconBackplane and support the virtual channel concept. The combined memory-SiliconBackplane system can deliver predictable, high-rate data streams to cores allowing end-to-end guarantees to be made around this service. However, DRAM systems are incapable of providing high-bandwidth guarantees for random accesses and are only efficient with block accesses. Block-addressing rules apply to data streams and affect the nature of performance guarantees. The DRAM controller is responsible for turning multiple data streams into interleaved block accesses to maximize the DRAM system's efficiency.

Dataflow Analysis

To map the application to SOC hardware, a dataflow graph of all required real-time functions was defined. The graph describes functional blocks, buffering requirements, and the dataflows between blocks with their worst-case bandwidths. Partitioning the system is accomplished by mapping this graph onto processors, memory, communications resources, and time.

For instance, the MPEG functions map onto DSP time slices for audio and special purpose processors for video. In places the MPEG data paths fold onto the SiliconBackplane, which connects functional blocks with the buffering resource, the DRAM subsystem.

The resulting mapped graph defines which dataflows the Silicon-Backplane will handle and the bandwidth requirements of each one.

Figure 3.8 shows one MPEG decode process that includes video and audio generation. The buffering required between processing stages is explicitly drawn and the bandwidths annotated (buffering required by the application rather than its hardware instantiation). The dotted boxes indicate how the MPEG function is partitioned into execution units. The dataflows crossing dotted boundaries represent interunit communications that need to be carried by the SiliconBackplane.

A complete dataflow graph for the STB audiovisual processing is shown in Figure 3.9. Two MPEG decoders, an MPEG encoder, VoIP, and other functions are included. It captures all real-time application functions, dataflows, and buffering. In the STB, most data buffering is handled by DRAM as shown in Figure 3.10, and all

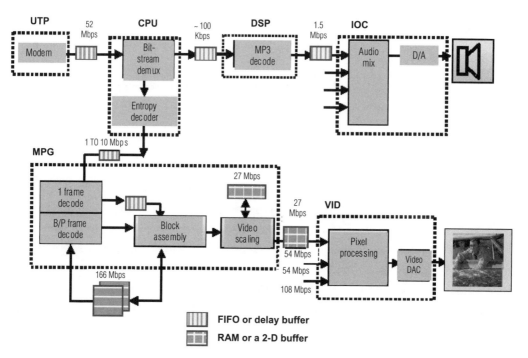

Fig. 3.8 MPEG Dataflow

3.5 SET-TOP-BOX SOC EXAMPLE

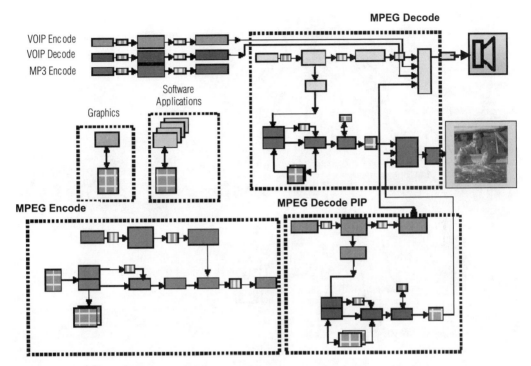

Fig. 3.9 Complete STB Dataflow

interblock flows are buffered. Redrawing the dataflow graph with a unified memory replacing the discrete FIFO illustrates the shared memory problem.

System Partitioning

The system architecture follows a conventional decomposition that includes a RISC and a variety of application-specific processors, I/O components, and a unified SDRAM subsystem connected by a SiliconBackplane. Partitioning is necessary since no single, general-purpose processor can achieve the performance levels required.

In the architectural definition phase, the cores and Silicon-Backplane infrastructure don't need to be concretely defined. For a variety of reasons, system requirements are likely to be highly fluid during development. This highlights the value of flexibility and the

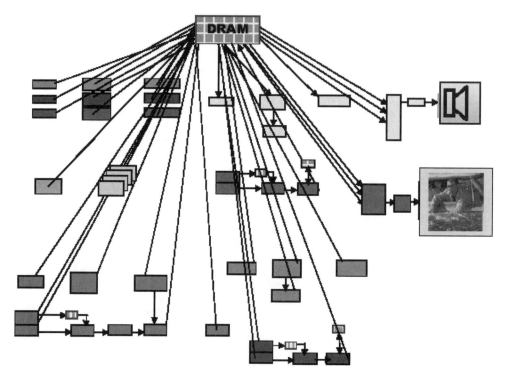

Fig. 3.10 Dataflow Buffer Mapping to DRAM

ability to rapidly upgrade. Since the Sonics SiliconBackplane allows late binding decisions on the integration and configuration of cores, it is suited for this type of application.

Block Descriptions

This section describes the function, implementation, and system requirements of the cores. Each real-time dataflow is identified and expressed in terms of bandwidth needs. In SiliconBackplane terminology, each dataflow is called a thread and each thread in the dataflow model is typically unidirectional, either reading from or writing to DRAM. For example, the video core will take two MPEG-generated images and a graphics plane on separate threads and combine the pixel streams into a single output. Video input from an analog source comes into the same core and out to DRAM on its own thread. Each thread may have different natural word sizes and data rates.

SDRAM Controller (SDC)

The SDRAM controller needs to provide hard, real-time bandwidth guarantees, minimum latency, and near-peak SDRAM utilization. The projected implementation uses double-data-rate synchronous DRAMs at 75MHz and a 64-bit-wide data path with 1.2GB/s peak bandwidth. The targeted peak bandwidth utilization is above 80 percent.

External SDRAM

The target devices are JEDEC standard (PC200), 4-page, DDR-SDRAM with a 4M word deep × 16b wide DDR-SDRAM. Four of these chips provide a 64-bit interface with a total RAM of 32MB. Using 128Mb devices as a build-time option provides the same performance and double the memory capacity. Other configurations are possible with the SDC providing similar performance and guarantees.

Video I/O Device (VID)

The video I/O device captures video from a camera or analog broadcast for compression and display (time shifting) or videophone type applications. A loopback path between video in and video out is provided for system test. The video output section takes constant rate pixel streams and manipulates them for display.

The pixel processor performs color-space conversion (from YUV and YCbCr to RGB) and the video overlay function. Graphics and MPEG pictures are mixed according to the alpha channel of the overlay video page. This provides video in a window, PIP, and full graphics display under software control. The video field rate was doubled from 60Hz to 120Hz to provide a flicker-free double-scanned display and is accounted for in the bandwidth requirements.

High-Speed RISC (CPU)

The central processor of the STB is used for system coordination and control (running RTOS), plus more specific real-time multimedia tasks such as handling parts of the MPEG encode process, MPEG data stream demultiplexing, or entropy coding.

In this application, CPU performance is optimized although it is not involved in many critical real-time flows. Real-time CPU processes can easily be overallocated to meet hard guarantees about their execution rates. Good non-real-time performance is valuable in many general-purpose applications like Internet browsing.

MPEG Processor (MPG)

The MPEG core needs strong, high-bandwidth guarantees from the memory system to decode video in real time. Compared to RISC, the processing can be deeply pipelined with lax latency requirements. Various threads are used by the MPEG core including compressed input stream, decoded video output, I and B frame macro block accesses, and some for MPEG encode. The core accesses memory as blocks corresponding to 16×16 pixels (a macro block), and the memory controller guarantees a number of these blocks per second. The pixel groups in MPEG decoding can straddle several blocks so a bandwidth assignment of four times the underlying raw rate is provided. This absolute worst-case provision will rarely be used, and the surplus SDRAM bandwidth can be applied to CPU or graphics operations. The MPEG encoder translates MPEG format, 4:2:0 YCbCr to 4:2:2, easing video output pixel processing since no line buffering is required for the video I/O device. This requires a little more bandwidth from the memory subsystem. This unit also handles picture scaling for picture-in-picture functions. An input bandwidth of MPEG1 (1.5Mb/s) or MPEG2 (up to 10Mb/s from DVD source or statistical multiplexed source material) is assumed. Overall, the bandwidth assigned to the MPEG unit is conservative and can be optimized. While a worst-case model is used, a supporting external environment can simplify the core design.

Digital Signal Processor (DSP)

The DSP implements audio-processing tasks including AC3, MP3 decode and audio encode, and a V.90 modem. Data and program caches are explicitly managed by the DSP so data access can be handled as block-burst transfers for code segments and sequential-block transfers for data. A simple estimate is sufficient for peak DSP-bandwidth requirements.

Utopia Interface (UTP)

A UTOPIA_II interface attaches to one or more cable modems or transceivers and supports xDSL, STV, and CTV including open-cable support. The interface can operate at up to 52Mbit/s. The CPU de-interleaves multichannel connections to provide 1.5Mbit/s to 10Mbit/s MPEG streams. These transceivers may be built on-chip or off-chip—a decision that can be made when better silicon area budget information is available. While this is a low bandwidth peripheral, audio and video quality depends on its integrity, so it requires performance guarantees.

Comms/Interfaces (IOC)

The IOC serves as a catch-all for input-output functions and its implementation is centered around a very fast, 8-bit RISC MPU core. These processors can implement in software what would normally be hardwired functions (even 10bT Ethernet) and provide a flexible way to meld multiple interfaces into one standard, flexible block. For instance, serialization of audio data for communication to an AC97 CODEC can be achieved, or a PC card interface synthesized from the MPU parallel ports and simple peripheral functions. Like the DSP, IOC threads are accounted for in the bandwidth requirements as an estimated, lumped bandwidth.

2D Graphics Accelerator (GPH)

The graphics controller is a very high bandwidth device that can saturate the SDRAM and SiliconBackplane. The 2D-graphics accelerator is conceived as a conventional bit_BLT engine. Generally the graphics controller's performance is not required to be real time. The controller is used in simulation as an initiator with high bandwidth demand to test the vulnerability of real-time processes to other high-speed processes.

Worst-Case Dataflow Definition

Table 3.3 defines the bandwidth needs of each core with real-time flows. The spreadsheet tabulates worst-case bandwidths where two MPEG decodes and an MPEG encode are occurring simultaneously with DSP program memory swap-out, maximum bandwidth I/O, and video refresh with picture-in-picture. Component dataflows are assessed in the spreadsheet and their bandwidths added to produce a required bandwidth number for each critical core.

The bandwidth requirements are exact for the video output section and Utopia interface, but conservatively estimated for the DSP and IOC sections. The video bandwidths account for peak, not averages that include blanking intervals.

Traditional Approach

A traditional video output section, shown in Figure 3.11, is closely coupled to DMA and DRAM controllers. The CPU may also control these blocks in a real-time, interrupt-driven manner, so designing the video block requires knowledge of the memory subsystem, DMA device, and other components. A specification for this core includes

3.5 SET-TOP-BOX SOC EXAMPLE

Table 3.3 Core Real-Time Bandwidth Requirements

	SB freq: 150 MHz	SB width: 64 bits			SB bandwidth: 1200 Mbps		
	Functional Unit	Data Format	Peak BW Mbps	Utilization of SB BW%	SB time wheel max spacing	Notes	
MPG	MPEG decode video o/p	YCrCb 4:2:2	54.00	4.5%	22.22	Channels I&II CCIR061 rate	
	MPEG decode MC	YCrCb 4:2:2	165.89	13.82%	7.23	MC: 4:2:2 for processing efficiency ñ absolute worst-case BW	
	MPEG encode video i/p	YCrCb 4:2:2	27.00	2.25%	44.44	Motion compensation I/B ref access	
	MPEG encode ME	YCrCb 4:2:2	82.94	6.91%	14.47	Motion estimation reference frame access	
	Encoded bit streams	MPEG	3.75	0.31%	320.00	3 streams	
	MPEG totals		333.58	27.80%	3.60		
VID	MPEG I out	YCrCb 4:2:2	54.00	4.5%	22.22	CCIR601, PAL or NTSC	
	MPEG II out	YCrCb 4:2:2	54.00	4.5%	22.22	—Double scanned for reduced flicker	
	Video graphics out	RGB	108.00	9.00%	11.11		
	Video/camera in	YUV to YCrCb	27.00	2.25%	44.44		
	VID totals		243.00	20.25%	4.94		
DSP	Audio in	PCM stereo	0.18	0.01%	6,802.72		
	Audio out	5-ch surround	0.44	0.04%	2,721.09		
	DSP program swap		256.00	21.33%	4.69	Peak rate for program memory swap	
	DSP totals		243.00	20.25%	4.94		
UTP	DSS etc in	Bit/byte stream	6.50	0.54%	184.62	Streaming data	
	XDSL etc out	Bit/byte stream	0.19	0.02%	6,400.00	Packet data	
	UTP totals		6.69	0.56%	179.44		
IOC	IOC data	Bytes, words	16.00	1.33%	75.00	Block access for cache line fills and spills, words, bytes in random access patterns	
	IOC program	Words	16.00	1.33%	75.00	Block access for cache line fills	
	IOC totals		32.00	2.67%	37.50		
	Overall Totals		**871.89**	**72.66%**			

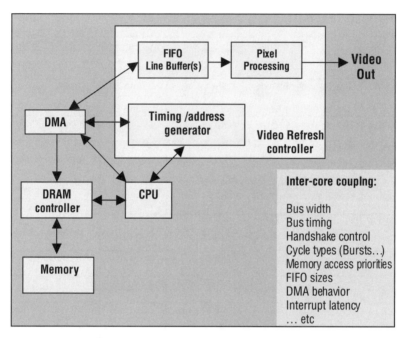

Fig. 3.11 Traditional Video Output Section

substantial detail, and the logic design and verification must cover all of the cores involved. The DMA designer, memory controller designer, and CPU integrator must also understand the video section since there is a high level of coupling.

Overdesign, particularly the use of large and expensive FIFOs, is a common technique used to reduce dataflow coupling. Buffer size depends on the characteristics of DRAM, its controller, and numerous other factors. Another form of overdesign is to employ private interfaces between critical blocks and the memory controller to provide individualized quality of service to different cores or to guarantee accessibility to memory by critical cores. Core designers are forced to consider architectural issues far beyond their core's boundaries, leading to errors and design management problems. A small change made to core timing or interface characteristics can drastically affect other cores. Without high-level functional decoupling, reusability of the cores is seriously compromised.

Sonics Approach

In the proposed STB design, the Open Core Protocol (OCP) used by the SiliconBackplane, the real-time facilities, and clock domain separation all provide the system with high-level decoupling. See Figure 3.12.

The interface on the memory side of the core is most efficiently sized to the natural word size of the video stream (larger only if there is a bandwidth limitation). The SiliconBackplane provides word length conversion and data packing so that the core designer need not know the communications path width used by other cores or within the SiliconBackplane.

Because the SiliconBackplane guarantees bandwidth availability, FIFOs are not required on the video I/O core. The responsibilities of the video I/O device designer can end at the OCP interface with-

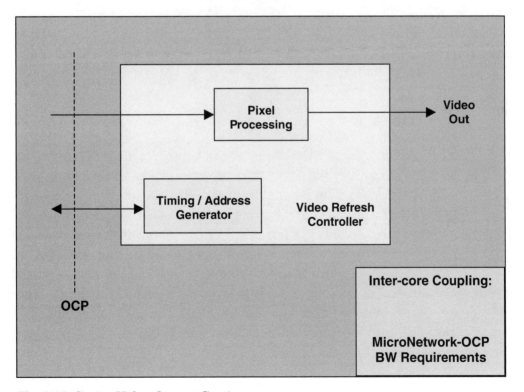

Fig. 3.12 Sonics Video-Output Section

out concern for the implementation of the DRAM controller. This decoupling greatly simplifies the design process and increases predictability, testability, and reusability.

The video I/O device core can be unit tested, probably before any DRAM models are available. Changes to the DRAM or communications systems do not need regression testing of attached modules providing that performance guarantees are maintained. One example could be doubling the SiliconBackplane data width and increasing its clock rate in a new application to get better performance from a new RISC. If the bandwidth assigned to the video section is maintained (now a much lower percentage of the SiliconBackplane traffic), the core need not be redesigned at all. Similarly, if the SDRAM controller is implemented with a different DRAM technology, the changes affect only the DRAM controller and not the interconnect or video hardware.

System integration, modeling, and simulation of the STB were performed using Sonics EDA tools. Also, the SiliconBackplane was configured to match DRAM bandwidth capability with the same data width of 64 bits and a frequency of 150 MHz.

SiliconBackplane

The SiliconBackplane MicroNetwork is configured to match DRAM bandwidth capability with the same data width of 64 bits and a frequency of 150 MHz. The SiliconBackplane TDMA time wheel, responsible for scheduling and arbitration control, is configured to provide the bandwidth needed by each initiator. RISC performance depends primarily on memory latency. A feature of the arbitration mechanism of the SiliconBackplane MicroNetwork and the prospective DRAM controller is that threads can be given priority until hard real-time deadlines are reached. This means that CPU latency can be kept close to the minimum SDRAM latency (even when contending with many hard real-time threads) providing the selected SDRAM bandwidth is not nearing saturation. CPU requests are serviced with minimum latency and very little degradation in performance even while sharing the MicroNetwork and memory with other devices. The typical bandwidth used by all the multimedia threads is much lower than peak requirements (for instance, MPEG infrequently requires use of its full allocation) and this extra bandwidth is available to the CPU, graphics controller, or other processes that benefit from it. This is another strong argument in favor of unified memory and resource sharing.

Modeling Methodology

Behavioral models are used to model the cores in the design and are configured to match the communications behavior of the cores running the target applications.

The focus of the model is the communications throughput of the SiliconBackplane. Simulation uses behavioral initiator or master models to generate worst-case traffic profiles. The system is deemed correct if the bandwidths demanded by the MPG, UTP, IOC, DSP, and VID cores are met by the system under all simulation conditions. A single slave or target model is used to simulate the SDRAM subsystem.

A mix of constant-rate flows (hard real-time), stochastic (CPU), and time-variant flows (typical of MPEG) were used as simulation stimuli. The CPU module generates a pseudorandom mix of burst read and word write transactions. The GPH module is used to saturate the SiliconBackplane by issuing continuous bursts at near-maximum rates at different periods in the simulation. Each hard real-time initiator issues a continuous burst during simulation with only the MPEG flow rate changing over time.

Dynamic load conditions were selected to highlight differences among the three arbitration types and to test functionality under complex flow conditions. In particular, sporadic high-bandwidth demands from one initiator must not interfere with the critical real-time processes of other cores.

Bandwidth Demand Model

The requesting rates over time are shown in Figure 3.13. For large portions of the simulation, demand exceeds peak memory bandwidth, which is typical in a system with multiple processors.

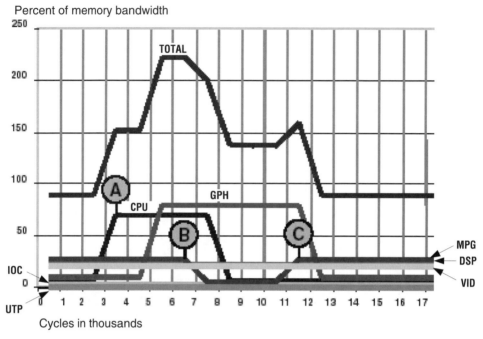

Fig. 3.13 Bandwidth Demand

Figure 3.13 shows the demand curves of individual initiators and total demand. The demand curves start with peak worst-case, real-time dataflows at around 72 percent of available system bandwidth. At point A, required bandwidth exceeds 100 percent with rising CPU demand, after which GPH demand rises to near peak bandwidth. At point B, MPG bandwidth use drops off. At point C, MPG demand returns. Initiators with unsatisfied demands accumulate transactions and post them as soon as possible at peak rates.

The pending transaction may take some time to complete depending on the actual bandwidth an initiator receives. After point C, CPU and GPH demands do not fall during simulation because they still have work in progress buffered up.

DRAM Model

A simple behavioral model is used for the DRAM subsystem that has an initial access latency and a subsequent burst latency for individual threads. In simulations, the latency was set to be 10 clock cycles for first access of a burst after which the memory subsystem can deliver or consume data at full bandwidth.

One limitation of the DRAM model is that latencies are modeled without bank contention, refresh, and other typical SDRAM timing constraints. This produces a bandwidth efficiency that is somewhat optimistic leading to a higher-than-expected Silicon-Backplane load.

Simulation

For the results, runs over 20,000 SiliconBackplane clock cycles (approximately 130µs) were simulated. Design iterations including compilation, VCS simulation, and sysperf analysis took less than one hour.

Results

Figures 3.14, 3.15, and 3.16 show the achieved bandwidth over time for each core. After point A, all results show very close to 100 percent system usage for the remaining portion of the simulation. The three arbitration schemes produce very different allocations of bandwidth.

Fixed-Priority Results

Figure 3.14 shows that a fixed priority can guarantee bandwidth to real-time initiators. However, the CPU absorbs all spare bandwidth and the GPH is starved and gets no bandwidth at all for a long period. With fixed-priority schemes, a danger is that deadlock can occur if one initiator swamps the system resources.

This simulation illustrates the vulnerability of any set of dataflows where the sum of bandwidth used by an initiator and all higher-priority cores exceeds system-bandwidth capability: extra bus and memory bandwidth is necessary to satisfy requesters before low-priority initiators get serviced.

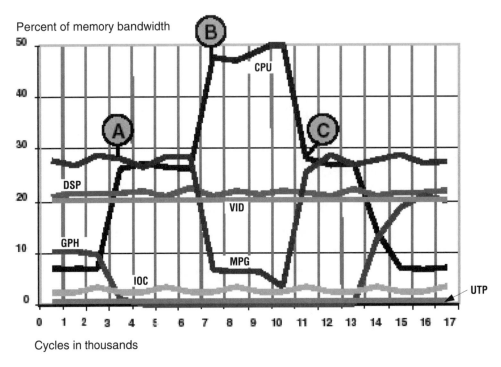

Fig. 3.14 Fixed-Priority Arbitration

Round-Robin Priority Results

Figure 3.15 shows the MPG falling below 20 percent and failing to receive its required bandwidth after the CPU increases demand at point A. The DSP and VID also fail to get required bandwidth after point B. Between points B and C, bandwidth is proportioned equally among the high-bandwidth initiators even though not all are critical.

The customary solution to this problem is overdesign of bus and memory bandwidth; however, real-time dataflows are vulnerable to overdemand from any other dataflows.

SiliconBackplane Results

For the SiliconBackplane, shown in Figure 3.16, hard real-time data delivery corresponds exactly to demanded bandwidth. The system keeps up with demand even at 100 percent bandwidth usage levels and is performing as designed.

3.5 SET-TOP-BOX SOC EXAMPLE

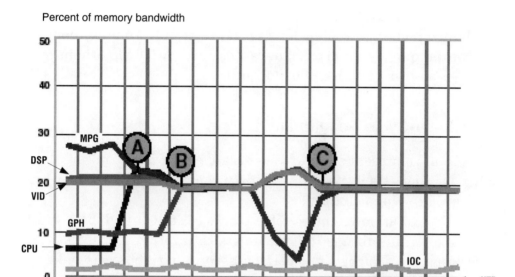

Fig. 3.15 Round-Robin Arbitration

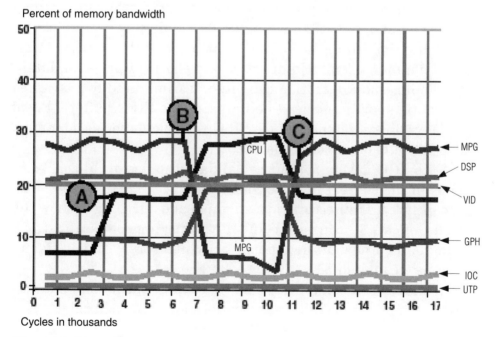

Fig. 3.16 SiliconBackplane MicroNetwork

The CPU and graphics share spare bandwidth, with the CPU taking a higher percentage due to additional TDMA allocation. Differential quality of service guarantees are satisfied without overdesign or overallocation of bandwidth.

Cycle-by-Cycle Results and Bursts

The graphs present a picture of operation over thousands of cycles, but looking at a cycle-by-cycle picture of performance shows the same effects in microcosm.

Typical initiators burst fetch blocks of data at high bandwidth. It is desirable to use longer bursts to improve DRAM system performance. During short periods, several initiators can make full bandwidth requests overloading the system in the same way, or worse, as in the cases previously described. Increasing burst length makes matters worse. To overcome resource contention and the resulting long worst-case latencies on this time scale, FIFOs and look-ahead accessing have to be implemented for real-time cores. FIFO sizing depends on many factors including the burst size of other cores, and, as such, undermines decoupling.

Using the SiliconBackplane arbitration with its finely interleaved dataflows can yield dramatically shorter worst-case latencies. Timing and arbitration can often be configured to guarantee sufficiently short latencies to eliminate core FIFO buffers. Fixed and deterministic latency makes FIFO size calculation easier and is independent of other core behavior.

This SOC design set out to validate the use of SiliconBackplanes in the context of a next-generation STB SOC design. The scope included architectural investigation and modeling to show the efficacy of the SiliconBackplane as a unifying communications medium and a new unified memory architecture with end-to-end performance guarantees.

While satisfying hard real-time bandwidth and latency guarantees in simulation, overall system performance was controllable and predictable and also indicated steady performance for cores such as RISCs under heavy traffic conditions.

Traditional shared-resource arbitration schemes make system performance dependent on the good behavior of initiators. A fair round-robin arbitration scheme relies on the requestors being fair; fixed-priority schemes also rely 100 percent on a trust model of one core on higher-priority cores. The SiliconBackplane does not rely on trust and uses a cooperative arbitration mechanism with enforced rules about bandwidth guarantees. Not only does this make the system more efficient, but application mapping is simpler and unexpected core behavior is easier to debug.

The SiliconBackplane greatly reduces data buffering requirements. Data buffers can be centralized in the core on which they depend and to which they are logically coupled, not distributed throughout the system as a patchwork solution to limitations in the communications infrastructure.

The performance of the SiliconBackplane in providing differential quality of service to many real-time data streams results in an almost perfect bandwidth efficiency. Computer-style buses without thread support need significant overprovision of bandwidth to reach the same levels of real-time performance. For systems without threaded data streams, memory bandwidth must be overprovisioned, or arbitrarily large buffering used to guarantee real-time functionality.

3.6 SUMMARY

In this chapter, we completed the two SOC examples (VoIP and STB) we started in Chapter 1. We also discussed design for integration.

In addition to the verification issues that we face when designing ASICs, we have new challenges in verification of SOCs that arise mainly from the mixing of silicon IPs. Verification planning and execution, automation, and IP verification are all key factors in SOC verification.

3.7 REFERENCES

1. J. Bergeron. *Writing Testbenches.* Norwell, MA: Kluwer Academic Publishers, 2000.

2. *www.cis.ohiostate.edu/~harrold/research/regression_testing.htm*
3. "Software Configuration Management with Elego ComPact." *www.elego.de/compact/concept-automation.html*, December 2000.
4. J. Plank. "Scripts and Utilities." *www.cs.utk.edu/~plank/plank/classes/cs494/notes*
5. M. A. Hernandez. "Creating and Using Makefiles." *http://csc.uis.edu/resources/makefiles.html*
6. M. Keating and P. Bricaud. *Reuse Methodology Manual for System-on-a-Chip Designs.* Norwell, MA: Kluwer Academic Publishers, 1999.
7. F. Nekoogar. *Timing Verification of Application Specific Circuits (ASICs).* Upper Saddle River, NJ: Prentice Hall PTR, 1999.
8. P. Rashinkar, P. Paterson, and L. Singh. *System-on-a-Chip Verification: Methodology and Techniques.* Norwell, MA: Kluwer Academic Publishers, 2001.
9. H. Chang and L. Cooke. *Surviving the SOC Revolution: A Guide to Platform-Based Designs.* Norwell, MA: Kluwer Academic Publishers, July 1999.
10. Spring Tech Seminars, Verification 2001, Synopsys, Inc., Mountain View, CA.
11. S. Furber. *ARM System-on-Chip Architecture*, 2d edition. Reading, MA: Addison-Wesley, 2000.
12. R. Michale and A. Betker. "Processor SuperCore Approach to Designing Complex SOCs." Lucent Technologies, 2000 IP World Forum.
13. Inrinsix Corp. "Multi-Platform, Mixed HDL, SOC Verification Environment." 2000 System-on-a-Chip Design Conference, Westboro, MA.
14. S. Azimi. "Overcoming Challenges and Obstacles to System-on-a-Chip (SOC) Products." Marvell Semiconductor, Inc., Sunnyvale, CA..
15. C. Settles (LSI Logic Corporation). "Silicon Development Platform Simplifies System Design." 2000 System-on-a-Chip Design Conference, Milpitas, CA.
16. D. A. Burgoon, E. W. Powell, L. J. Sorensen, and J. A. S. Waitz (Hewlett-Packard Company). "Next-Generation Concurrent Engineering for Complex Dual-Platform Subsystem Design." 2000 System-on-a-Chip Design Conference, Santa Clara, CA.
17. D. Tavana and S. K. Knapp (Triscend). "A Configurable System-on-a-Chip Device Facilitates Customization and Reuse." DesignCon 2000, Santa Clara, CA.

CHAPTER 4

Physical Design

4.1 INTRODUCTION

In this chapter, we discuss the physical design that is commonly called "back-end" design. Regardless of the type of IC design you have, your physical design techniques apply to both traditional ASICs and modern SOCs. The same basic principles and techniques are applied for floorplanning, placement, routing, physical verification, timing optimization, etc. How you plan for your physical design in advance is the key to a successful physical design.

We assume that the reader is familiar with the basic concepts of physical design. Therefore, we concentrate on the modern design techniques and cover some tips and guidelines for hierarchical physical design.

Section 4.2 reviews the physical design flow. Here, we review the basics of floorplanning, place and route, clock-tree implementation, timing optimization, and physical verification.

Section 4.3 introduces some tips and guidelines for hierarchical designs.

Section 4.4 covers two examples of modern physical design techniques.

4.2 OVERVIEW OF PHYSICAL DESIGN FLOW

We reviewed front-end design process in Chapter 1. Figure 4.1 shows a simplified front-end design flow. Here we included floorplanning as part of the front-end design flow. However, many consider floorplanning as a step in the back-end design flow. Whether it is part of front-end or back-end design flow, the main purpose of the floorplanning is to provide the location of blocks, I/Os, and power pads, as well as to define clock and power distributions. In order to do floorplanning, you need to know the gate size and operating frequency of each block in your design.

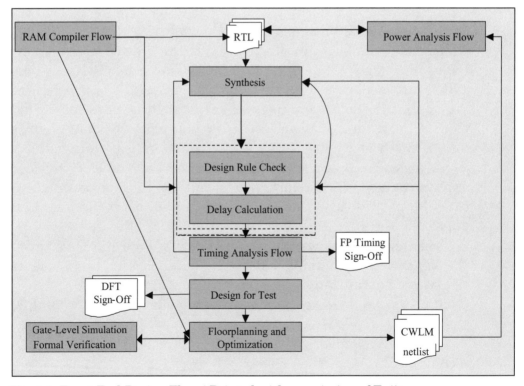

Fig. 4.1 Front-End Design Flow (Printed with permission of Fujitsu Microelectronics America, Inc.)

4.2 OVERVIEW OF PHYSICAL DESIGN FLOW

Figure 4.2 shows a simplified back-end design flow. Here, we start with a final floorplan which predicts the size of the chip; then we proceed with timing-driven placement. Timing-driven placement allows for timing optimization based on SDF.

Before planning for your clock insertion, you should know the answers for the following questions:

- How are the (core, reference, source, and recovered) clocks distributed?
- What are the core-clock and recovered-clock frequencies?
- What is the reference-clock frequency?
- What is the reference-clock jitter requirement? How is the reference-clock buffering?

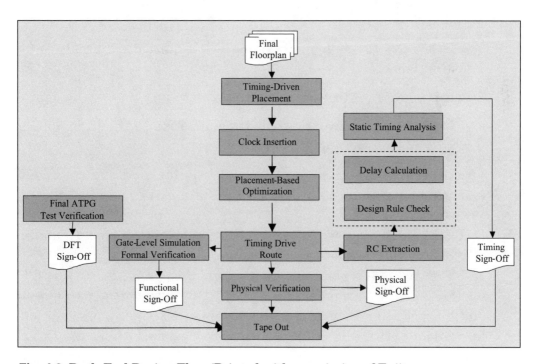

Fig. 4.2 Back-End Design Flow (Printed with permission of Fujitsu Microelectronics America, Inc.)

- ☞ What are the receiving and sending source-clock frequencies?
- ☞ What is the jitter required for sending clock?
- ☞ What is the jitter on distributed clock?

Clock distribution is one of the most critical aspects of chip design. Criteria for clock distribution are skew and insertion delay, and keeping two values as low as possible. There are several ways to implement clock distribution.

Clock-tree synthesis is a method of implementing clock distribution that generally achieves small skew with relatively low cost (short implementation time with no special cells). However, it requires aid from CAD tools because the location of distributed clock buffers is critical for skew control. The drawback of this method is that clock insertion delay increases in large chips.

Mesh-based clocking is another scheme which has the following benefits:

- ☞ It is easy to implement, though it requires a unique clock-buffer macro.
- ☞ It is easy to control skew.
- ☞ It does not require any netlist modification to the logic netlist after placement and routing.

However, the disadvantages of this method are required resources for generating clock macros, large power and metal consumptions.

EDA tools are used to generate balanced clock tree. Cadence CTgen is an example of a clock-tree tool.

The goal of the place-and-route stage is to produce a GDSII file for tapeout and also meet timing constraints. The input to place-and-route is I/O specification, gate-level netlist, timing information for critical paths, and clock description. Basically, there are three main steps in routing:

- ☞ Global routing

4.2 OVERVIEW OF PHYSICAL DESIGN FLOW

- Detailed routing

- Fixing violations

Global routing provides a plan for reducing the critical path delays as well as the length of interconnects throughout the design. This information plus a congestion map (to show placement routability) are made available for the detailed routing stage to route every net. However, the routing of special nets such as clock and power nets is usually performed before detailed routing.

Detailed routing completes the connections between logic cells and reduces the total logic area and critical path delays throughout the design based on the routing information from global router. The major goal of this stage is to accomplish all of the required interconnects. Channel routing and area-based routing are examples of of detailed routing.

Placement, global routing, detailed routing, and fixing routing violations such as shorts and spacing violations, delay calculation, and clock-tree synthesis are the major tasks performed by place-and-route tools. One popular EDA routing tool is NanoRoute from Cadence Design Systems. NanoRoute performs global and detailed routing, as well as post-routing optimization for any cell-based, block-based, or mixed cell- and block-based designs to reduce the chip-level integration effort. It also fixes the design violations with post-routing optimization to improve the TAT.

Timing optimization can be divided into two parts: pre-route prediction and post-route correction. Examples of optimization techniques are gate sizing, buffer insertion/deletion, and placement optimization. Figure 4.3 illustrates some of these techniques. PhysicalStudio from Sequence Design, Inc., is an example of a timing-optimization tool.

In general, Signal Integrity (SI) issues consist of power problems (voltage drop and electromigration) and timing (crosstalk) problems. Voltage drop or IR drop refers to the drop in the supply voltage across the power distribution network. This affects storage

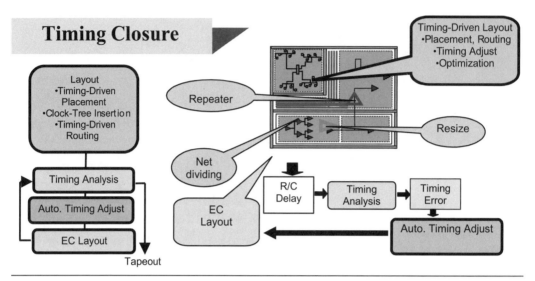

Fig. 4.3 Timing Optimization (Printed with permission of Fujitsu Microelectronics America, Inc.)

capabilities of RAMs, flip flops, and other storage elements as well as performance and timing failure.

Metal electromigration is a problem that occurs on power wires caused by increased wire resistance and stronger electric fields on finer geometries. These can produce short circuit on adjacent metal lines and broken lines. Electromigration problems worsen with increased current densities and higher temperatures.

Crosstalk problems are caused when the cross-coupling capacitance between two adjacent nets adversely affects the signal integrity of one of the nets. This can result in serious timing problems. Cadence Silicon Ensemble is an example of an EDA tool for crosstalk analysis that you run after routing.

All signal integrity issues have to be resolved for a successful tapeout.

Physical verification consists of the following steps:

☞ Design Rule Check (DRC) is necessary to guarantee that a circuit can be fabricated with an acceptable yield. Design rules ensure that the design will not fail due to shorts or process faults.

- ☞ Layout versus Schematic (LVS) compares electrical circuits (source and layout) from the specified sources.
- ☞ Antenna Check ensures that induced capacitance is under a certain value so that electrical characteristics stay stable.
- ☞ Electrical Rule Check (ERC) operations perform tasks related to electrical rule checking that have to be followed by circuit designers. ASIC vendors provide these rules. Examples of ERC violation are open input, short circuit, NMOS connected to Vdd, and PMOS connected to Gnd.

Two popular EDA tools for physical verification are Mentor Calibre and Cadence Dracula.

In order to have a successful tape out, the following sign-off procedures must be complete:

- ☞ Functional sign-off
- ☞ Timing sign-off
- ☞ DFT sign-off
- ☞ Physical sign-off

4.3 SOME TIPS AND GUIDELINES FOR PHYSICAL DESIGN

The following are some guidelines for hierarchical design structures. For designs larger than two million gates, some hierarchical plus partitioned design methodology should be used. The fact that parallel processing is performed makes hierarchical methodology an attractive implementation for physical designs.

Use Placement-Based Synthesis

As we discussed in Chapter 2 for designs larger than two million gates, a placement-based synthesis approach must be taken. Figure 4.4 shows the design flow for a placement-based synthesis.

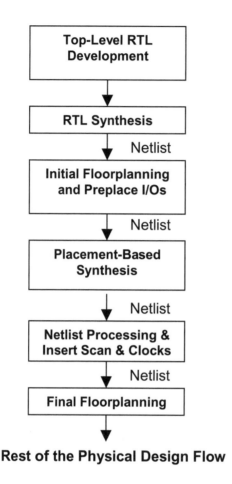

Fig. 4.4 Design Flow with Placement-Based Synthesis

Logical versus Physical Hierarchy

In a hierarchical design, initially you can have an unlimited number of hierarchies in your logic design. However, after generating a floor-plan with a chip-level timing budget, logical hierarchy should match the physical hierarchy. Setting the timing budgets for each block in your design and making sure that you haven't overconstrained certain blocks while underconstraining others are important points to consider.

4.3 SOME TIPS AND GUIDELINES FOR PHYSICAL DESIGN

Figure 4.5 shows that the logical hierarchy matches the physical hierarchy. Here hierarchical layout block (HLB) is a block that can be laid out independently as a hard macro.

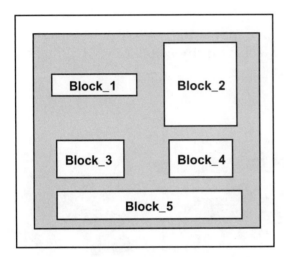

Logical Hierarchy

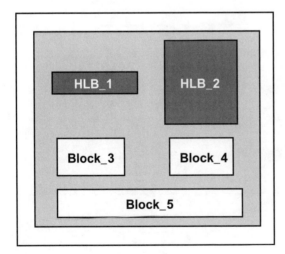

Physical Hierarchy

Fig. 4.5 Logical versus Physical Hierarchy (Printed with permission of Fujitsu Microelectronics America, Inc.)

Clock Design

Figure 4.6 shows a two-level clock-tree design where you have chip-level and block-level clocks. Apply the following for your clock implementation:

1. **Chip-level clock tree**
   Designers should prepare only one clock-tree block at the Level_0. This level of hierarchy consists of HLB, clock block, and repeater cell.

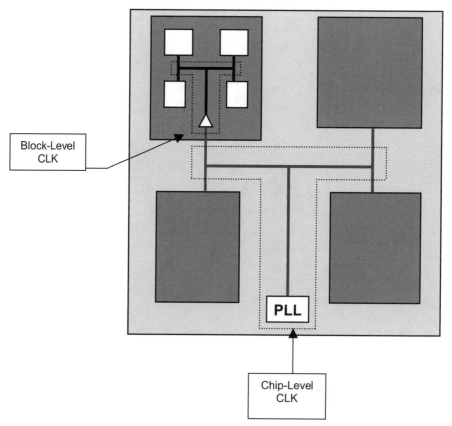

Fig. 4.6 Two-Level Clock-Tree Design (Printed with permission of Fujitsu Microelectronics America, Inc.)

2. **Block-level clock tree**
   Designers should prepare only one clock-tree block at the Level_1 (block level). Do not distribute the block-level clock to other blocks.
3. **Clock balancing with upper and lower blocks**
   Designers should take care of clock balancing between Level_0 and Level_1.

Global Bus Design

Figure 4.7 shows an example of a global bus design.

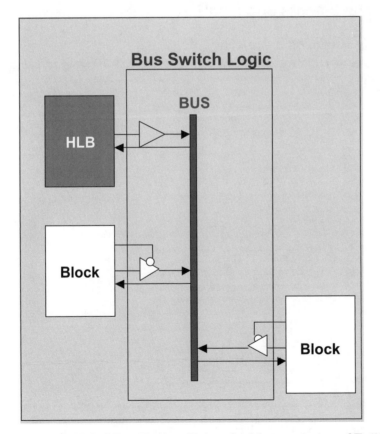

Fig. 4.7 Global Bus Design (Printed with permission of Fujitsu Microelectronics America, Inc.)

1. Logic designers must control the bus-switch logic based on the floorplan.
2. Do not use tri-state buffers for chip-level bus, for the following reasons:

   Cannot insert a repeater cell

   Cannot drive a high capacitance bus load

3. Allowing for bus switch logic placement, you should resynthesis or redesign the bus switch logic as well as the bus-arbitration logic.

Multiple Placements and Routing

When using hierarchical methodology, it is best to use multiple placements and routing, as shown in Figure 4.8. The first place and

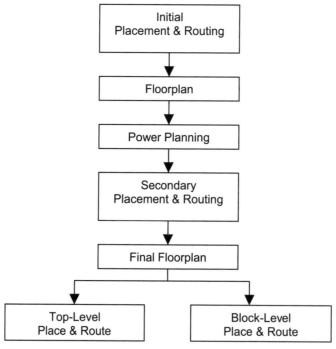

Fig. 4.8 Multiple Placements and Routing in a Hierarchical Methodology

route will require an estimate for the location of your blocks. All subsequent placement and routing stages are based on fixed-block assignments and adjusting I/O positions.

Nonroutable Congested Areas

Apply the following when you experience a nonroutable congestion in your design:

1. Change the aspect ratio.
2. Change the distance between the blocks/macros.
3. Allow the routing over the block.
4. Change the location of the block/macro.
5. Go back to the initial placement again.

4.4 MODERN PHYSICAL DESIGN TECHNIQUES

In the age of deep submicron design, where 10+ million gates of logic have to fit on a single device running at 250+ MHz, traditional physical-design techniques are not capable of handling these new challenges. The problems with the traditional physical-design techniques can be summarized as follows:

- Timing closure is either unachievable or takes too long to finish.
- Too many iterations between front end and back end for each design.
- Unroutable designs for the target die size.

As the device geometries shrink to 0.11 micron and beyond, new tools, techniques, and methodologies are needed to overcome the problems we face with traditional approaches.

Here in this section, we cover two examples of modern physical-design techniques. The new methods presented here both overcome the above challenges and substantially simplify the physical design.

The Silicon Virtual Prototype[1]

In the mid-1990s, two important issues were converging to make physical chip designers' lives miserable. First, they were being asked to put more and more on a single chip. The integration challenge really hit home as chip design approached the million-gate hurdle and physical-design tools choked on the complexity. Second, the interconnect started to have a much stronger effect on the timing for the entire chip. Timing closure became a huge issue because most physical-design tools did not comprehend the effects of the interconnect.

As a result of these two issues, chip physical implementation became a huge problem, especially as geometries shrank to 0.25 microns and below. Logic designers were told that their designs simply couldn't meet timing. Physical designers were dismayed at the lengthy run times of their choking tools, especially when many iterations were required to converge on a physical implementation that could meet all the timing, area, and power requirements.

The existing methodology promoted an arm's-length working relationship between the front-end logic-design team and the back-end physical-layout team. Each group used completely different tools, geared specifically for its use only. This methodology relied on multiple iterations between the front and the back ends. As a result, new chip designs were significantly delayed.

What did designers do? There seemed to be two choices. One was to resort to overly pessimistic designs with wider than necessary guardband as a safety margin. However, this flew in the face of the demands for higher and higher performance. The other choice was to suffer the pain of 20 or more iterations between synthesis and layout, if they ever brought about a timing closure at all.

---

1. Courtesy of Silicon Perspective, Inc. (A Cadence Company). Portions reprinted with permission.

What was needed was a tool/methodology combination that effectively bridged the gap between the front-end logic design and the back end, where the tools were falling far short. This new technique needed to provide fast turnaround for quick feedback between the front end and back end instead of the days or weeks required to complete one layout iteration and generate the necessary timing and simulation data for feedback to the front-end process. What also was needed was a totally new approach that squeezed between the front end and back end. That new approach is what is now called the silicon virtual prototype, and is what has made it possible to design multimillion-gate chips today.

This section will discuss how First Encounter from Cadence's Silicon Perspective subsidiary can be used to create a silicon virtual prototype, helping designers to reduce iterations and finish large-chip designs much faster.

Silicon Virtual Prototyping The silicon virtual prototype is the practical approach to dealing with increased complexity and the challenges of timing closure with deep submicron (DSM) silicon. The silicon virtual prototype creation is the first stage of the back-end design phase, often before all of the front-end design is complete. By creating a full-chip silicon virtual prototype, the design team can immediately validate the physical feasibility of the netlist—eliminating the back-end design iterations that were required to discover that a chip could not meet timing or some other constraint. Figure 4.9 illustrates the environment for physical prototyping.

The silicon virtual-prototyping stage compresses the physical-feasibility stage down to a few hours so chip designers can view the chip layout once or more each day. Now designers can evaluate many implementations of their designs and get quick feedback on the best trade-offs. Key to the speed is the revolutionary change in tool-database architectures. Design files for traditional tools tended to be very large, limiting the maximum design size they could effectively handle. The new tools employ lightweight databases that let them handle multimillion-gate designs with ease.

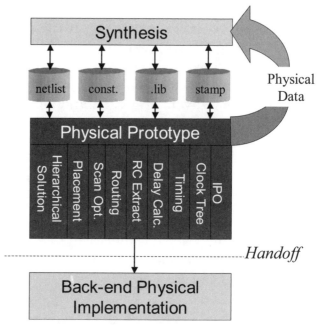

Fig. 4.9 Physical Prototype Environment

The creation of the prototype also allows the designers to create realistic timing budgets for all sections of the chip. These timing budgets can be set based on real physical information from the actual design rather than an estimate.

The New Design Flow In this new methodology, designers may start building their silicon virtual prototype at a very early stage in a chip's development process. For a very large design, the prototype creation can begin when portions of the design are incomplete. Black boxes can be used to estimate those regions that are not complete or consist of third-party IP.

Figure 4.10 shows the design flow for silicon virtual prototyping. Once the functionality of the chip is fully defined, the first step in this methodology is to perform a quick logic synthesis to create a gate-level netlist. It is assumed that the netlist is functionally clean but that timing is not accurate, so simple wire-load models (WLMs) can be used at this stage. The resulting gate-level netlist plus the timing constraints form the inputs to the creation of the silicon virtual prototype.

The creation of the silicon virtual prototype begins with fundamental full-chip floorplanning activities such as I/O placement, macro placement, and power topology. Major design elements usually are manually placed based on the designer's knowledge of the chip architecture. The remaining elements can be automatically placed for maximum efficiency.

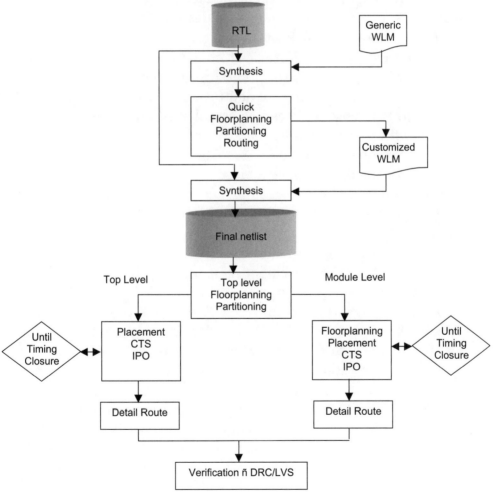

Fig. 4.10 Design Flow for Silicon Virtual Prototyping

Conventional floorplanning tools start by creating block shapes and fitting them together, then use this template to try to drive a full layout. Floorplanning for a silicon virtual prototype works in almost exactly the opposite way. It creates a full, flat placement of all standard cells and macros for the entire chip. Silicon Perspective's First Encounter tool uses an algorithm (called *Amoeba*) to generate a fast placement that is timing driven and intelligently blends logical hierarchy and physical locality to optimize concurrently for timing, power, die size, and routability. The Amoeba clustering technology preserves the logic hierarchy on the physical side so changes can be traced through the design.

For the first pass at floorplanning, a netlist, physical libraries, corresponding synthesis libraries (.lib), top-level constraints, and a technology file (process description) are created and imported. This data can be automatically loaded for succeeding passes. Specific macro and I/O placements can be established and then saved into a command file so they can be read in for later iterations. Placement guidelines can be generated and saved to indicate, for example, that certain cells need to be placed close to the I/Os. Placement guides usually are created for the major modules and used to guide the placement engine as where to roughly place the module cells. A script for power/ground topology also can be created.

Next, the remaining standard cells are placed using a timing-driven algorithm. The placement includes a trial route that ensures that major congestion issues are eliminated. The design is then extracted and the timing analyzed. This gives the design team a quick idea of the physical feasibility of their design and allows them to experiment with different placements to determine the most desirable aspect ratio and optimum block locations.

Timing Closure Conventional approaches to timing-driven layout typically rely on slack-based, constraints-driven algorithms. These are "explicit" timing-driven methods in that they depend on prespecified timing constraints to direct the layout operation. These techniques work relatively well when the design is small, but with any design of reasonable complexity, the algorithms bog down and performance deteriorates rapidly. In addition to being very hard to use and slow to run, these conventional approaches often achieve timing closure at the expense of layout quality in the areas of wire length, die size, and routability.

In contrast, the Amoeba placement engine takes an "implicit" approach to timing control. Throughout the entire floorplanning and placement process, it tries to exploit the natural clustering characteristic that is intrinsic in a design's logical hierarchy. Instead of depending on externally imposed constraints, the Amoeba engine naturally blends logical hierarchy and physical locality in its core algorithmic foundation. By exploiting the logical hierarchy, physical locality is achieved naturally, in tandem with optimizing wire length, timing performance, chip die size, and layout routability.

The Amoeba engine applies this hierarchical-locality-based approach to a unified floorplanning and placement task. At any given level of a typical hierarchical chip design, intramodule signals account for over 95 percent of all signal nets, leaving fewer than five percent for intermodule signals. The physical locality implicitly leads to shorter wire lengths for intramodule signals, which generally require smaller drivers. Exploiting this characteristic allows the Amoeba technology to focus on the intermodule signals, which are greatly reduced in number and tend to be more critical. The Amoeba engine applies techniques such as net weighing, soft and hard planning guides, grouping/regrouping, and rehierarchy. Additionally, hard placement fences are employed, interwoven with other features such as power/ground strips and obstructions.

The Amoeba technology only needs to deal with a very drastically reduced number of signal paths at any given stage in the placement process because it applies the physical locality and intramodule/intermodule signal distribution hierarchically and incrementally. This greatly enhances its speed and allows it to more thoroughly explore the possible solution space to develop an optimum solution considering all the factors of timing, area, and power consumption.

The Amoeba technology uses an intelligent-fencing strategy to place circuit cells. Conventional placement approaches confine cells in a design module to a nonoverlapping "fenced" rectangular or rectilinear area. Such rigidity leads to wasted die area and greatly impaired routability. With the intelligent-fencing strategy, Amoeba allows two clusters to overlap when and if necessary, leading to more efficient die usage, shorter wire lengths, and much better routability.

One of the big benefits of the Amoeba-based approach is that it helps designers make intelligent decisions about allocating timing budgets among different blocks. Designers can easily see if certain parts of the chip are overconstrained or underconstrained and make the appropriate adjustments. Here, the budgets are set based on real physical data.

To verify timing closure, designers compare the timing data produced by the prototype against the final tape-out version of the chip. By comparing the Standard Delay Format (SDF) files produced by an extraction and timing analysis of both versions of the design, the design team can make sure timing goals are met. Correlation results for this type of analysis should yield 90 percent of the nets to be within 10 percent.

Design Partitioning and Hierarchical Techniques This new methodology mixes the best of both hierarchical and flat design. By first creating a totally flat global view of the entire design and evaluating different physical placement options (in the step described above), designers get the best of flat design techniques. However, most tools choke on multimillion-gate flat designs. Therefore, a top-level partitioning can be employed to optimally break the chip into a number of hierarchical blocks for physical implementation. The hierarchical approach also makes it possible to have multiple design teams working in parallel on different sections of the chip, speeding the completion of the design.

Hierarchical methodologies have been widely adopted in the front-end logical design world. However, designers have hesitated to embrace hierarchical methodologies for physical design because of

4.4 MODERN PHYSICAL DESIGN TECHNIQUES

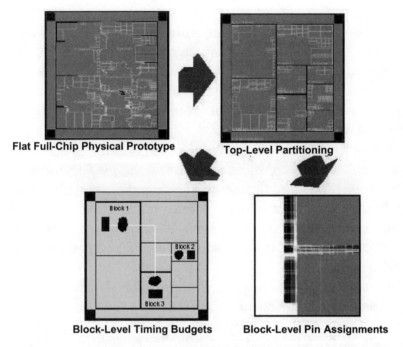

Fig. 4.11 The Full-Chip Physical Prototype Drives the Partitioning

the challenge of generating accurate timing budgets and pin placements for the blocks. First Encounter provides the intelligence designers need to allocate timing budgets among the blocks and to determine optimal pin placements. Figure 4.11 shows how the full-chip physical prototyping drives partitioning.

The silicon virtual prototype is the starting point for creating a physical hierarchy in the design. During the import, all modules are flattened to create the prototype. The standard cells are placed flat at the top; the design is then routed and extracted and the timing is analyzed. This is when the partitioning is implemented to re-create the hierarchy.

The tool creates a directory of data for each partition, including the top level. Each directory contains a netlist, floorplan file, pin assignments, and timing constraints. In addition, the subdirectory for the top cell contains a block view for each partition.

Reaching the optimal block size in large designs often requires two levels of partitioning. The size of these sub-blocks is driven by the capacity of the tools, such as physical synthesis, that seem to perform best on blocks of 100,000 gates. For example, a design of five million gates would be partitioned into 10 blocks of 500,000 gates. Those blocks would then be partitioned into sub-blocks of approximately 100,000 gates.

To be able to perform the second-level partitioning, the constraints that were created as a result of the partitioning must be combined with the multicycle and false-path constraints for each partition. As with the major partitions, placement, routing, extraction, and timing analysis will derive the necessary timing budgets for the second-level partitions, which then can be passed to a physical-synthesis tool.

Automatic Pin Assignment First Encounter can produce an automatic, optimized assignment of pin locations on chip partitions, using both detailed logical and physical information. This eliminates a time-consuming, tedious, and error-prone task.

Block-Level Physical Synthesis, Place, and Route Physical synthesis tools can be used on each block of the hierarchical design. Additionally, block-level place-and-route tools, such as Cadence Silicon Ensemble-PKS or Synopsys' Physical Compiler, are used at the block level. As each block is completed, it is placed back into the silicon virtual prototype to make sure the design is on target for timing, area, and power.

Chip-Level Implementation for Hierarchical Design As the design team creates the physical blocks, they are assembled back into the silicon virtual prototype. However, a number of top-level or chip-level tasks are also required. The master clock network for the chip must be generated, the power and ground framework must be designed, and the top-level interconnect and buffering must be created. First Encounter provides advanced capabilities that supplement the design team's existing router for these tasks.

A key element of chip-level assembly is managing buffers between the design's partitions in order to achieve top-level timing closure. First Encounter works with popular commercial routers to provide a fast, flexible mechanism based on either rules or timing. First Encounter performs top-level buffer insertion when the prototype is created. However, in practical design flows, minor netlist changes continue to be made within the chip's blocks well into the physical design stage; often these changes have implications for top-level chip timing and even routing.

In-place optimization (IPO) downsizes, upsizes, and inserts buffers and repeaters to obtain an optimal timing once all the block designs are complete. IPO needs to be performed at the top level to guarantee timing closure at the partition boundaries.

The next step is clock-tree synthesis. First Encounter's clock-tree synthesis option creates a complete clock tree at both the top level and the block level, leveraging the Amoeba-placement algorithm to minimize skew and insertion delay. Even very complex clock structures can be handled.

In hierarchical design flows, most design teams use automated tools to generate clock structures within the chip's partitions. However, top-level clock trees usually are created manually. First Encounter's clock-tree synthesis provides automation at both levels, leveraging the Amoeba ultra-fast placement technology to generate and optimize a top-level buffer-tree network that balances the clock-phase delay for minimum skew between the chip's partitions. This results in major time savings over manual methods.

First Encounter's clock-tree synthesis supports gated clocks for use in power-sensitive applications and generates a clock-routing guide for final detailed routing.

Power Grid Design Power distribution has become a significant concern in SOC designs. Most designers now routinely over-design their chip power grids in order to prevent IR-drop problems. However, not only does this interfere with signal routing (congestion), but it also increasingly does not guarantee that IR-drop violations will be avoided.

The First Encounter Power Grid Designer option addresses this problem. It enables designers to lay out and analyze the chip's power and ground network early in the physical design cycle, with accurate correlation to the final layout. This detects potential IR-drop problems early in physical design, instead of at the end of the layout cycle. Power Grid Designer leverages First Encounter's trial route, extraction, and delay calculation engines to deliver results that previously required an entire physical layout cycle to achieve. This, in turn, allows designers to optimize the power grid while the cost of correcting problems is still low.

Benefits of the Silicon Virtual Prototype One of the key benefits of developing a silicon virtual prototype is the turnaround speed of each iteration. In a traditional design flow, information about physical feasibility is available only after the completion of place and route followed by final verification. This process typically takes several days. A silicon virtual prototype tool must be able to validate the physical feasibility of a large SOC design in a few hours on a desktop workstation.

There are numerous other benefits as well. The silicon virtual prototype provides a vehicle for much better and increased communication between the front-end designers and the back-end designers. Design teams can try new approaches and quickly find out if they're on the right track or if they are wasting their time. Above all, the silicon virtual prototype provides an efficient way to speed designs through the physical design cycle. First Encounter provides a production-ready placement that is ready for final routing.

New Approach for Implementing Big IC Design[2]

Shrinking device size is driving increased gate count in ICs, but the design methodology is not keeping up with the pace. Most design houses have EDA tools and hardware that can run flat designs up

---

2. Courtesy of AmmoCore Technology, Inc. Portions reprinted with permission.

to two million gates or hierarchical designs up to four million gates. Most of the hierarchical design methodology uses block-based design, with each block size averaging 200,000 to 500,000 gates.

There are no alternatives to designs over two million gates, but block-based designs have several rigid constraints both in terms of timing as well as physical real estate in a given die.

Block-Based Design As shown in Figure 4.12, the block-based design methodology imposes artificial boundary constraints. It is very difficult to meet the critical timing path requirements. While it is easier to meet the block-level timing constraints, it is very difficult to meet chip-level timing requirements. This causes endless iterations to close chip-level timing requirements. In most cases, the chip speed is sacrificed to meet the market window.

Several EDA vendors have come up with floorplanning solutions, hoping to solve the mysteries of timing closure, but they all failed because early gross estimation is as good as no estimation.

The challenge is how to arrive at a 98 percent confidence level within a few hours, not in a few days. Achieving even 70 percent is not good enough, which is the claim of most floorplan vendors.

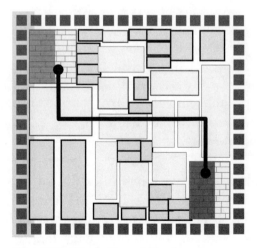

Fig. 4.12 Block-Based Design

New Approach The new approach, as shown in Figure 4.13, is to break down these rigid boundaries of physical blocks and come up with faster, more fluid solutions. Ammocore Technology, Inc., has come up with a unique approach that allows designers to minimize their floor-planning task, break their design into smaller groups, and process all these groups via parallel processes. This methodology allows designers to gather meaningful data within a few hours without hitting any design size (gate count) limitations, and it offers proof that a design of four million gates can be implemented within 14 hours from netlist to GDSII.

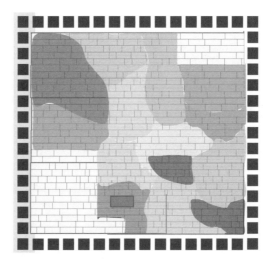

Fig. 4.13 New Approach to Physical Design

Design Flow There are five major steps in the implementation for this new approach. The steps are floorplanning, partitioning, placement, assembly, and verification. Figure 4.14 shows the design flow for this new approach.

Floorplanning In this basic floorplanning step, designers place the hard macros (e.g., RAM, ROM, IP core) and implement power for such macros. There are no requirements for doing month-long floorplan optimizations based on wire-load models. Typically, this can be done within an hour for a large design.

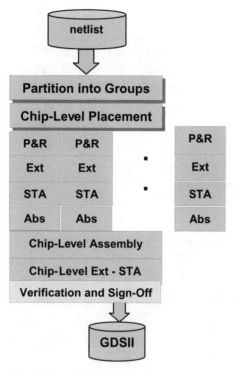

Fig. 4.14 Design Flow

Partitioning The partitioning step not only considers connectivity and timing but also looks at the available spaces in the die. This step breaks the design into manageable pieces and is the key to meet timing and routing requirements in fewer iterations. All these pieces are then implemented using parallel processing, before proceeding with the top-level placement.

Placement Placing all the partitioned pieces at the chip level provides significant insight into the likelihood of meeting both timing and routing requirements. If there are issues in timing or routing, designers can fix them by inserting buffers or optimizing logic, or even by programming minor perturbation of the placement. The time it takes to do partitioning and placement is about an hour for a design of up to four million gates.

Assembly Assembly happens in two steps, first on the small pieces and then on the chip level. Assembly time is significantly reduced due to the nature of two-step process: Designers can debug any timing issues related to critical paths faster by using these two steps. As a result, the router needs to evaluate a smaller percentage of nets at the top level.

Verification Verifying a large design is always tricky. Many design houses struggle to come up with a streamlined process for timing and physical verification before submitting the design to the mask shop. Our approach is to divide the problem into a more manageable size and then process the pieces in a massive parallel fashion. In this way, most of the issues will be resolved before the design is sent for a final verification using traditional tools that takes days, if not weeks.

In summary, traditionally, block-based design methodology is used for large designs, but it imposes many constraints on a designer. Block-based methodology requires several iterations at the block level as well as at the chip level. In many cases, it is very difficult to achieve the performance goals due to the nature of rigid boundaries in the die.

The new approach breaks these boundaries and provides an innovative way to handle design of any gate count. Managing the data and processing it in a massively parallel way reduce debugging time to a ratio of 1:100. The two-step process also improves the design implementation time to a ratio of 1:10.

4.5 SUMMARY

For a successful tape out, you should complete the following steps:

- ☞ Functional sign-off
- ☞ Timing sign-off
- ☞ DFT sign-off
- ☞ Physical sign-off

Some tips and guidelines for physical design were discussed in Section 4.3 including the following:

- ☞ After generating a floorplan with a chip-level timing budget, the logical hierarchy should match the physical one.
- ☞ For hierarchical methodology, use multiple placement-and-routing stages.

Examples of optimization techniques are gate sizing, buffer insertion/deletion, and placement optimization.

Two examples of new physical design techniques were presented in Section 4.4.

4.6 REFERENCES

1. F. Nekoogar. *Timing Verification of Application-Specific Circuits (ASICs)*. Upper Saddle River, NJ: Prentice Hall PTR, 1999.
2. P. Rashinkar, P. Paterson, and L. Singh. *System-on-a-Chip Verification Methodology and Techniques*. Norwell, MA: Kluwer Academic Publishers, 2001.
3. *Physical Studio/ShowTime Reference Manual*. Sequence Design, Inc., 2002.
4. M. J. S. Smith. *Application-Specific Integrated Circuits*. Reading, MA: Addison-Wesley, 1997.
5. S. Chow, A. B. Kahng, and M. Sarrafzadeh. "Modern Physical Design: Algorithm, Technology, Methodology (Part III)." ICCAD Tutorial, November 2000.
6. *Envisia Silicon Ensemble Place and Route*, Training Manual, Version 5.3. Cadence Design Systems, Inc.
7. *First Encounter*, Silicon Perspective Corporation (A Cadence Company) data sheet.
8. ASIC Products Application Notes, *Application of Synopsys Physical Compiler in IBM ASIC Methodology*, IBM, August 2001.
9. ASIC Products Application Notes, *Application of Cadence Envisia PKS in IBM ASIC Methodology*. IBM, May 2001.
10. B. Young. *Digital Signal Integrity: Modeling and Simulation with Interconnects and Packages*. Upper Saddle River, NJ: Prentice Hall PTR, 2001.

CHAPTER 5

Low-Power Design

5.1 INTRODUCTION

The current trend toward portable electronic devices, such as wireless multimedia systems that demand high speed for computations and complex functionality, has made the design of low-power integrated circuits more and more imperative. Since these devices run on small batteries, minimum power consumption and, therefore, increased battery life are very important for the consumers of such products. Reducing the power consumption in portable devices increases reliability as well as portability while lowering the cost of production. Therefore, the continuous need for effective methods of implementing successful low-power design techniques is a major challenge faced by ASIC/SOC designers. Traditionally, the major factors determining the design methodologies for chip designs were timing and size. Figure 5.1 shows three major independent factors influencing today's design methodologies. These factors are area, timing, and power. Here, it is assumed that other factors such as TTM and reuse have already been considered.

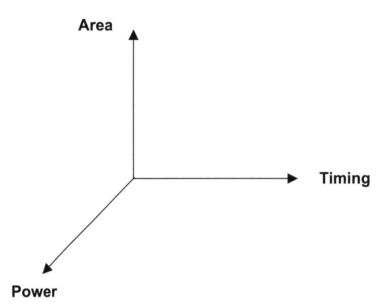

Fig. 5.1 Factors Affecting Design Methodologies

In Section 5.2, we define power in integrated circuits. Sources of power dissipation in CMOS devices are covered including static, short-circuit, and dynamic power dissipations.

In Section 5.3, we discuss low-power design techniques such as optimization techniques at algorithm and architectural levels. RT-level optimization techniques such as clock gating and memory-power reduction are also discussed.

In Section 5.4, we cover low-power design tools. This section should be used in conjunction with Appendix A.

Section 5.5 summarizes some tips and guidelines for designing low-power ASICs and SOCs.

5.2 POWER DISSIPATION

Power in electronic devices is defined as the conversion of electrical energy of power supply to heat. Equation 5.1 represents the power dissipation in electric circuits.

5.2 POWER DISSIPATION

$$P = V \cdot I \qquad \text{(Eq. 5.1)}$$

Where:

V = Voltage (Joules/Coulomb or Volts)

I = Current (Coulombs/Sec or Amperes)

P = Power (Joules/Sec or Watts)

Now let's look at dissipation in Complementary Metal Oxide Semiconductor (CMOS) devices. CMOS technology is the best choice for low-power designs because of its insignificant static power dissipation. However, simply selecting CMOS technology should not be considered as the only method for reducing power in ASIC/SOC devices.

Sources of Power Dissipation in CMOS Devices

Since most of today's designs are based on CMOS technology, the first step toward power reduction is to understand the sources of power dissipation in such devices. Power consumption sources in digital CMOS circuits are divided into three main categories:

- ☞ Static power dissipation
- ☞ Short-circuit power dissipation
- ☞ Dynamic power dissipation

Figure 5.2 shows a CMOS inverter with these three sources of power consumption.

Equation 5.2 illustrates the relationship between these three parameters.

$$P_{Average} = P_{Static} + P_{Dynamic} + P_{Short\ Circuit} \qquad \text{(Eq. 5.2)}$$

CMOS devices have very low-static power dissipation and most of the energy in them is used to charge and discharge load capacitances. Figure 5.3 shows an example of charging and discharging the load capacitance in a CMOS inverter.

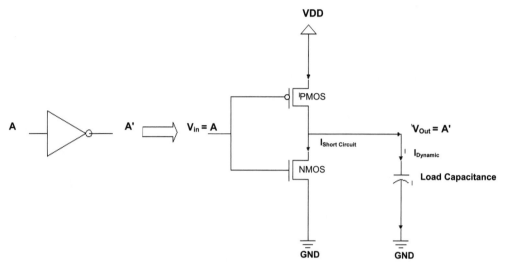

Fig. 5.2 CMOS Power-Dissipation Parameters in an Inverter

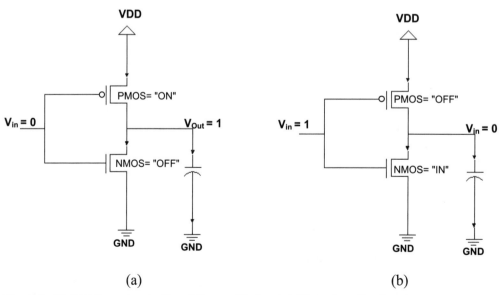

Fig. 5.3 CMOS Inverter in Two Binary States (a) Charging the Load Capacitance; (b) Discharging the Load Capacitance

As shown in Figure 5.3, the load capacitance in CMOS circuits is charged when positive voltage flows through PMOS transistor from V_{DD} to represent logic 1. Similarly, the load capacitance is discharged through NMOS transistor to represent logic 0.

By comparison, the short-circuit and static powers are usually of smaller magnitude than the dynamic power, and they can be ignored. Therefore, dynamic power is the principal source of power dissipation in CMOS devices. The following sections explain each of these power dissipation sources in detail.

Static Power Dissipation

Static power dissipation occurs when the logic-gate output is stable; thus it is frequency independent. Equation 5.3 represents the static power components.

$$P_{Static} = V_{DD} \cdot I_{Leakage} \qquad \text{(Eq. 5.3)}$$

Leakage current is caused by subthreshold-transistor operations and is determined by device technology. This type of current is responsible for power dissipation when a CMOS device is inactive and its value is insignificant (less than 1 percent) when the device is active. Therefore, the large amount of leakage current, or static power, accordingly is an indication of a serious design problem, such as static inputs that do not turn a gate on or off properly.

Static power dissipation in CMOS devices is usually negligible, because the amount of leakage current can be decreased significantly by choosing appropriate device technologies.

Short-Circuit Power Dissipation

As shown in Figure 5.3, short-circuit power dissipation occurs when current flows from power supply (V_{DD}) to ground (GND) during switching. The value of short-circuit dissipation depends on the amount of short-circuit current flowing to GND and it accounts for almost 10 percent of CMOS power consumption. Equation 5.4 represents the short-circuit power dissipation.

$$P_{\text{Short Circuit}} = V_{DD} \cdot I_{\text{Short Circuit}} \qquad (\text{Eq. 5.4})$$

Short-circuit current decreases when a large capacitive load is seen by the output of a gate and is at its maximum value when there is no capacitive load.

Dynamic Power Dissipation

Dynamic power is the dominant source of power dissipation in CMOS devices and accounts for approximately 90 percent of overall CMOS power consumption. It occurs during the switching of logic gates, and as a result, this type of power dissipation is frequency dependent. Dynamic power is therefore the average power required to perform all the switching events across the circuit. Equation 5.5 defines various parameters of dynamic power dissipation.

$$P_{\text{Dynamic}} = 1/2 \cdot \beta \cdot C \cdot V_{DD}^{2} \cdot F \qquad (\text{Eq. 5.5})$$

Where:

β = Switching Activity per Node

C = Switched Capacitance

F = Frequency (switching events per second)

V_{DD} = Supply Voltage

From Equation 5.5, it is evident that dynamic power can be reduced by lowering the supply voltage, switched capacitance, switching activity per node, or frequency of signal transitions from 0 to 1 or vice versa. It is also apparent from Equation 5.5 that the most effective and simple way of reducing dynamic power dissipation is by lowering the supply voltage (if the option of choosing lower voltage is available for a device). This is due to the squared effect of V_{DD}. The other three terms in Equation 5.5 influence the overall power dissipation linearly.

The switching activity (β) determines the amount of switching that occurs in each node. Lowering this parameter decreases the useless transitions. β can be estimated statistically or captured from simulation traces (for example in Verilog, a .vcd file).

Frequency represents the switching events per second. Since dynamic power is frequency dependent, frequency reduction is a key concept in power optimization. Clocks are the major contributors to the frequency component of Equation 5.5. However, other signals such as bus interconnect signals contribute to high-frequency activity and should be lowered to optimize power.

Switched capacitance (C) can be either estimated based upon statistical models or measured from an actual layout database. Switched capacitance can be lowered by using shorter interconnect wires and smaller devices.

5.3 LOW-POWER DESIGN TECHNIQUES AND METHODOLOGIES

Low-power techniques vary depending on the level of the design targeted, ranging from semiconductor technology to the higher levels of abstraction. These abstraction levels are classified as algorithm, architecture, RT, gate, and transistor levels. Figure 5.4 shows various levels of hierarchy that should be considered for low-power designs.

The higher levels of design abstraction shown in Figure 5.4 provide larger amounts of power reduction for chip designs. In higher levels of abstraction, such as algorithm level, designers have a greater degree of freedom to implement low-power design techniques. Hence, power-optimization process is the most effective method in higher levels of abstraction.

As we move toward the lower levels, the amount of power savings becomes less significant and the speed of power optimization becomes slower. Lower level power-optimization techniques are more accurate; however, they are not as fast as higher level meth-

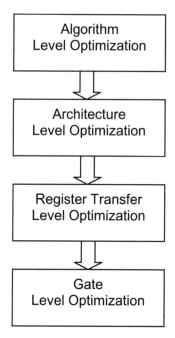

Fig. 5.4 Levels of Design Power Optimization

ods. Table 5.1 illustrates the amount of power savings for various optimization methods.

It's important to mention that a successful power-optimization methodology requires low-power techniques to be carefully considered at each level of design abstraction.

In previous chapters, we covered the design flow and methodologies for front-end and back-end designs. Figure 5.5 shows a typical

Table 5.1 Power-Saving Percentage per Optimization Method

Optimization Method	Power-Saving Percentage
Algorithm Level	75%
Architecture Level	50%–75%
Register Transfer Level	15%–50%
Gate Level	5%–15%
Transistor Level	3%–5%

5.3 Low-Power Design Techniques and Methodologies

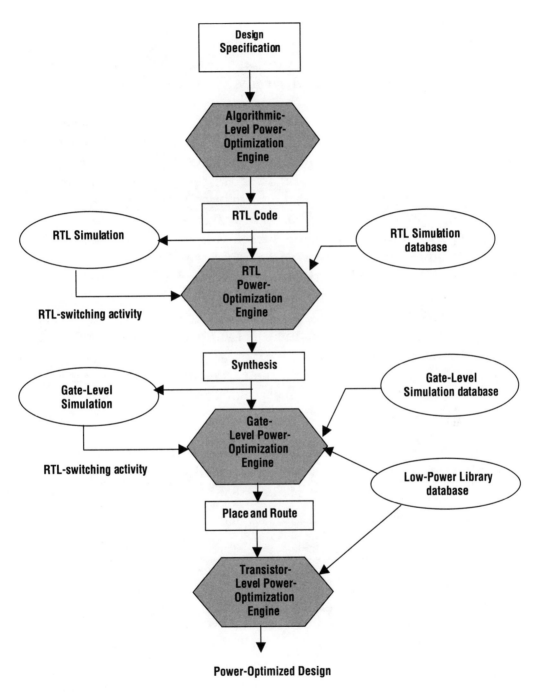

Fig. 5.5 Low-Power Design Flow

low-power design flow. In this figure, the RT-level and gate-level simulation databases and switching-activity data are input to the corresponding power-optimization engines. Here, the last stage in power optimization is performed at transistor level where a low-power library is used.

As mentioned in previous sections, short-circuit and static power dissipations have minimal effect on overall power consumption and can be minimized by selecting the appropriate physical components and process technologies. Therefore, dynamic power is the only source of dissipation that can be effectively managed through proper design techniques. In order to optimize power in a design successfully, dynamic power in each individual component (memories, interconnects, processors, cores, etc.) has to be estimated and minimized separately. Reducing the dynamic power requires the minimization of all parameters in Equation 5.5.

Table 5.2 is a brief summary of various factors involved in saving dynamic power.

The following sections cover several low-power techniques based on reduction of parameters in Table 5.2 at various levels of design abstraction. As mentioned earlier, higher levels of design abstraction contribute more to power savings than do the lower levels. However, power-reduction techniques have to be considered at all levels of design to obtain the best results.

Table 5.2 Factors in Reducing Dynamic Power

Reduction Parameter	Result
β	Saves power by optimizing the excessive transitions through clock gating
C	Saves power by using an appropriate process technology or improving the layout
V_{DD}	Saves power dramatically but slows down the circuit
F	Saves power through reducing the clock frequency but results in slowing down the functionality

Algorithm-Level Optimization

Algorithm-level optimization provides the highest level of power savings in a design. In this stage, hardware/software partitioning provides power reduction by dividing the tasks between hardware and software. Using HW/SW partitioning, heavy calculations can be performed in the hardware, while control functions can be handled in software. This partitioning reduces the chip's power consumption and minimizes the processor load and bus traffic. An example of HW/SW partitioning is shown in Figure 5.6 in a Voice over Network (VoN) application.

The heart of this SOC is a code excited linear predictive (CELP) core for voice compression/decompression. The CELP functions in this example are divided between hardware and software. Computationally intensive functions of CELP, such as codebook search and vector quantization, are performed in the hardware, while the system processor is used only for data formatting and control.

Algorithm-level optimization is commonly used in processor and DSP modules/cores. DSP algorithms are computationally intensive. Therefore, reducing the number of operations to execute a given DSP function lowers the switching activity and results in reduced power. Most DSP algorithms involve several multiplication operations. Multiplication, which is the most power consuming operation in such

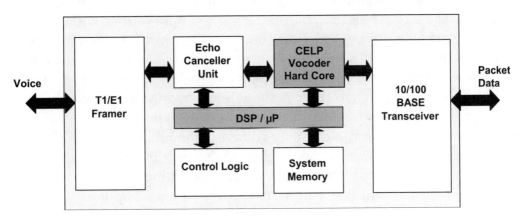

Fig. 5.6 HW/SW Partitioning in a VoN SOC

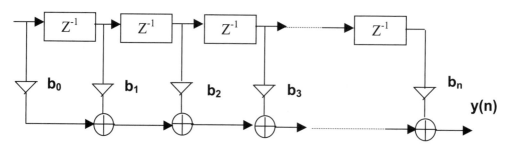

Fig. 5.7 General FIR-Filter Implementations

algorithms, can be replaced by shift-add operations to reduce power. For example, this method can be useful in finite impulse response (FIR) filter designs. Figure 5.7 represents an FIR filter implementation diagram.

As shown in Figure 5.7, in an FIR filter an array of constant tap coefficients is multiplied by an array of delayed data samples. The resulting array is summed with the most recent data samples after each multiplication. Multiplication by constant coefficients can be replaced by shift-add operations to lower the power consumption in such filters.

Architecture-Level Optimization

According to Equation 5.5, power is dependent on the supply voltage squared. Therefore, minimizing the supply voltage will result in remarkable savings in power consumption. One approach for power optimization using supply-voltage reduction is to use a common low-supply voltage for all of the logic modules. However, lowering the supply voltage increases the latency in various blocks and as a result decreases the overall speed of the design.

Utilizing multiple supply voltages is another approach to overcoming the problem of large delays that result in lowering the speed of a device. Using this method, each logic block, or each group of logic blocks, can use a unique supply voltage. Applying multiple supply voltages in a chip allows the modules on critical or high-speed paths to meet their timing requirements by using a high supply voltage, whereas the modules on noncritical or low-speed paths can use lower supply voltages. Hence, the power is saved without influencing the overall speed of the design.

The basic idea of minimizing the switching activity to reduce power can also be performed at the architectural level for processors by carefully using arithmetic representations. Two's complement is the most common representation used in DSP blocks/modules. This is because arithmetic operations such as additions and subtractions are easily performed with two's complement. However, the problem with this representation is that its sign extension causes the sign of the most significant bit (MSB) to toggle when signals change sign. To handle this problem, the use of sign-magnitude arithmetic-number representation is recommended. This approach can reduce the switching activity compared to two's complement numbering for negative values. The reason is that in two's complement numbering a large number of unnecessary ones is needed to represent a small negative number. Therefore, the number of bit transitions between positive and negative numbers is large and results in high-transition activity for input signals around zero. The following is an example of two's complement representation from −4 to +4.

Decimal	Two's Complement
−4	1100
−3	1101
−2	1110
−1	1111
0	0000
1	0001
2	0010
3	0011
4	0100

As shown in the above example, there is high activity in the shaded area around zero.

Another area where power can be minimized at architectural level is memory. We will cover memory-related power optimizations in later sections.

RT-Level Optimization

RTL power optimization reduces the high activity of signals through clock gating, finite-state-machine (FSM) encoding, bus encoding, and avoiding glitches. Each of these concepts will be covered in detail in the following subsections.

Clock Gating Optimization Clock signals in ICs are considered to be the major contributors to power dissipation because they switch at all times. Clocks can be gated to reduce excessive switching in synchronous registers. Figure 5.8 represents a 4-bit register with and without clock gating.

As shown in Figure 5.8b, the clock signal is transitioning continuously at the clock input of all four flip flops. However, when the clock signal is gated, it transitions only at input of the AND gate. As a result, the register with the gated-clock scheme dissipates less power than the one without clock gating. Using this method, the amount of power savings increases as the number of bits in a register increases. Gate-level power-optimization EDA tools automatically add gates to clocks in order to save power. Appendix A provides detailed information on some of the popular EDA power-optimization tools.

Signal-Encoding Optimization Power can be reduced in algorithm level by choosing the appropriate coding style. The use of gray codes results in a significant decrease in switching activity. Minimizing the number of transitions in the state assignment of an FSM has a considerable influence on lowering the power consumption. If the next state of an FSM differs by one variable from the preceding state (minimum Hamming distance) activities are minimized and

5.3 LOW-POWER DESIGN TECHNIQUES AND METHODOLOGIES

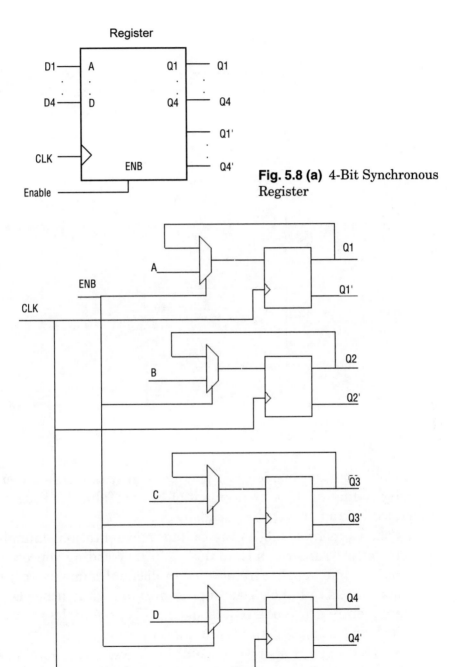

Fig. 5.8 (a) 4-Bit Synchronous Register

Fig. 5.8 (b) Internal Circuitry of a 4-Bit Register without Gated Clocks

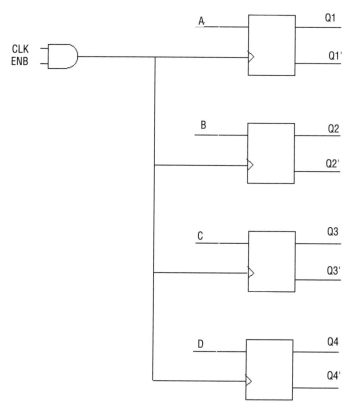

Fig. 5.8 (c) Internal Circuitry of a 4-Bit Register with Gated Clocks

power can be saved considerably. Therefore, gray coding is preferred to binary coding in state assignment of FSMs. Table 5.3 illustrates binary coding and gray-coding schemes.

As shown in Table 5.3, a gray-coding representation of numbers requires fewer transitions than does a binary-coding representation. For example, moving from 3 to 4 in decimal numbers requires three transitions in binary-code representation, while it needs only one transition in gray-code representation. Therefore, less power is consumed when states of an FSM use gray coding.

This technique is also very useful for power reduction in SOC bus interconnects. Buses are significant sources of power dissipation (almost 20 percent of total power dissipation in most SOCs are due to bus interconnects) because of their high switching activity and large

Table 5.3 Transition Activity of Binary Coding versus Gray Coding

Decimal	Binary Code	Number of Transitions	Gray Code	Number of Transitions
0	000	0	000	0
1	001	1	001	1
2	010	2	011	1
3	011	1	010	1
4	100	3	110	1
5	101	1	111	1
6	110	2	101	1
7	111	1	100	1

capacitive loading. Therefore, bus signal encoding is very effective in lowering the switching activity. Equation 5.6 shows how Equation 5.5 can be modified to represent power dissipation in bus interconnects.

$$P_{Dynamic} = 1/2 \cdot n \cdot \beta \cdot C \cdot V_{DD}^2 \cdot F \qquad (Eq.\ 5.6)$$

where n is the number of bits in a bus.

Since dynamic power is directly related to bus width, bus segmentation can provide considerable power reduction in interconnects by reducing capacitance in each bus segment.

As mentioned earlier, gray coding can provide less transition activity for bus interconnects. Another commonly used encoding technique is bus invert (BI) coding. In this method, a control signal determines if the actual data or its complement provides fewer transitions to be sent at each clock cycle on a bus interconnect. This decision is made based on the Hamming distance between the present and next state of the data bus. If the Hamming distance is larger than half the bus width (n/2), the next bus value will be equal to the complement of the next data value. If the Hamming distance is smaller than n/2, the bus value will be equal to actual data value.

Partial bus invert (PBI) coding can also be used to reduce the switching activity in interconnects. In this method, a wide bus is segmented to a smaller subset of buses, and the high activity subsets are encoded using bus inverting. All of the mentioned bus encoding techniques reduce the activity in interconnects. However, gray coding is commonly used for instruction buses, BI is used for data buses, and PBI is most common in address buses.

Combinatorial Transitions Optimization (Deglitching) As mentioned earlier, switching activity in CMOS devices is a major cause of power dissipation. Therefore, protecting circuits from unnecessary switching can save power significantly. Glitches are momentary transitions that occur in combinatorial circuits due to delay imbalances in different gates. Figure 5.9 illustrates an example of glitching in a logic block.

Since glitches add to the number of signal transitions, they should be avoided whenever possible. As explained in algorithmic level optimization, algorithms that require many multiply-and-add

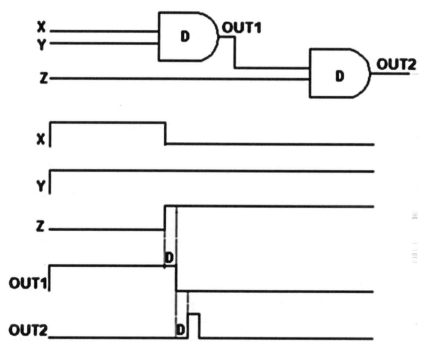

Fig. 5.9 Glitching in Logic Gates

operations, such as filters in general, are prone to produce glitches and consume a lot of power. These arithmetic operations can produce glitches if all of the multipliers and adders are sequential without any latches to hold their value until they become stable. Figure 5.10 represents two different implementations of the multiply-add operation for such algorithms.

In Figure 5.10a, depending on the delay of adders, the result of each add operation may have glitches until the final stable result occurs. Glitches can propagate to the next level of adders as inputs and then generate momentary wrong results, causing more transi-

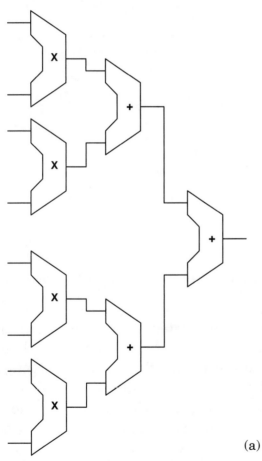

Fig. 5.10 Implementation of a Multiplier and Adder Tree (a) Non-Latch-Based Implementation

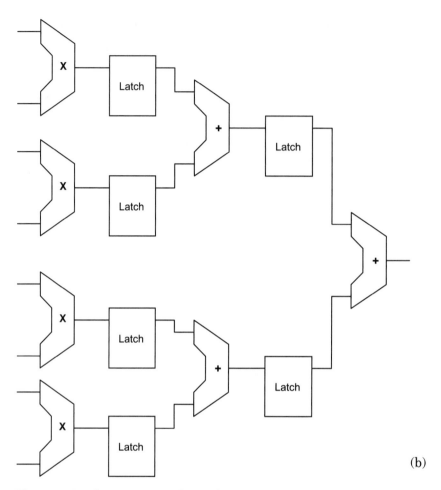

Fig. 5.10 Implementation of a Multiplier and Adder Tree (b) Latch-Based Implementation

tions until the final result is generated. These momentary transitions increase power consumption in the circuit and should be avoided by using latches at the output of each adder and multiplier, as shown in Figure 5.10b.

Similar to flip flops, latches save the previous value of the inputs at each level and prevent the extra switching activity that results in glitches and momentary wrong results.

Another low-power design technique that is very common among chip designers is to replace the flip flops with latches whenever possible. Both latches and flip flops are building blocks of sequential circuits and their outputs depend on the current inputs as well as previous inputs and outputs. Figures 5.11 and 5.12 illustrate the difference between a D-latch and a D-flip flop respectively.

As shown in Figures 5.11 and 5.12, the main difference between latches and flip flops is that latches are level sensitive to clock signal, while flip flops are edge sensitive. For example, in a D-latch, output Q obtains the value of input D for as long as the clock signal is asserted; thus any changes in input will be transferred to output as long as the clock signal is at a specific level (called transparent latch). However, in a D-flip flop, the output changes only at the edge of the clock signal. Therefore, in flip flops the input is

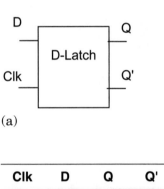

(a)

Clk	D	Q	Q'
1	1	1	0
1	0	0	1
0	1	0	1
0	0	0	1
1	1	1	0
1	0	0	1
0	1	0	1
0	0	0	1

(b)

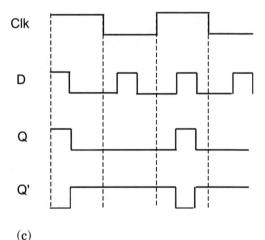

(c)

Fig. 5.11 Positive-Level Triggered D-Latch (a) Logic Symbol (b) Truth Table (c) Timing Diagram

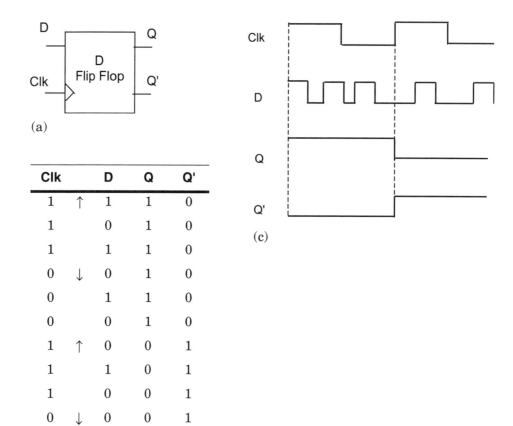

Fig. 5.12 Positive-Edge Triggered D-Flip Flop (a) Logic Symbol (b) Truth Table (c) Timing Diagram

transmitted to output only at a specific edge of the clock, while in latches the input is transmitted to output anytime the clock signal is at a specific level. After the rising or falling edge of the clock, the D-flip flop's output remains constant even if input changes.

Memory Optimization

Memories are considered a large factor (about 40 percent) of the total power dissipation in ASICs/SOCs. There are three types of power consumption in memory modules/cores:

- Power dissipation associated with program and data memory accesses
- Power required to transmit data across the large capacitance of system interconnects
- Power consumption within memory units

Memory accesses are power hungry because writing to and reading from memory locations requires high-switching activity on data and address buses. Therefore, in many DSP algorithms, such as image and voice processing, lowering the number of memory accesses required to complete a given function reduces the power consumption in such cores. By having more code in the cache, a designer can minimize the number of power-intensive external accesses to the main memory.

Memory can also be as a group of independent memory sections that can have their own individual clock signals. In this type of architecture, when each memory section is idle, its clock can stop transitioning and the memory segment can be put to sleep mode. Therefore, a large amount of power can be saved when various sections of memory are idle.

Another popular low-power practice among designers is to split large memories into smaller memory modules. For example, a 512K × 32 RAM can be split into four 128K × 32 RAMs and still provide the same total number of read cycles with lower power dissipation in smaller memories compared to the single large memory. Figure 5.13 illustrates the memory-partitioning approach.

Power Management

Power management is a technique that controls the power of a chip at various instances by switching off a core (sleep mode) when not needed. It can be done either in software or pure hardware. In a chip with multiple clock domains, various clocks can be gated and a control signal can be used to activate a specific clock only when the block is active. A clock-control block implemented in software or

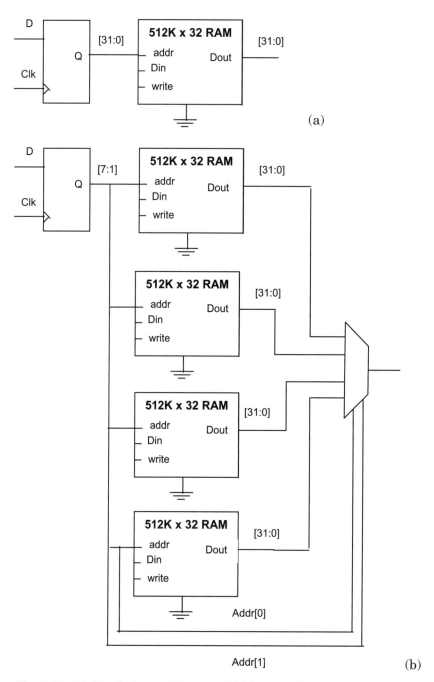

Fig. 5.13 (a) Single Large Memory (b) Memory Partitioning

5.3 LOW-POWER DESIGN TECHNIQUES AND METHODOLOGIES

hardware can provide different clock rates to manage various clock domains. To distribute the clock and control the clock skew, a clock tree needs to be constructed using clock buffers. In this approach, some portions of the chip can be shut down at various instances. Figure 5.14 is an example of a clock-control block that delivers separate clock signals to various soft and hard blocks.

In this technique, different clocks are derived from a master or central clock and can be slowed or stopped under certain conditions to avoid the unused switching activity. Therefore, the load on the master clock is reduced and the flip flops receiving the derived clocks are not triggered when they are on idle cycles. For example, in designing a global positioning system (GPS), several sets of filters per satellite can be controlled by separate clocks. This approach can significantly reduce power consumption in the design of such systems. Figure 5.15 is an example of multiple clock domains in a GPS.

Although gated clocks reduce excessive switching activity and hence the power consumption in synchronous circuits, they can create complexity in static timing verification of such designs. This is because of additional setup time constraint caused by the control signal. EDA STA tools, such as PrimeTime from Synopsys, have

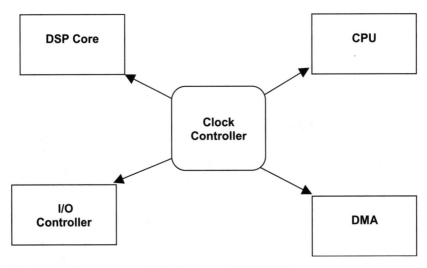

Fig. 5.14 Clock-Control Block in an ASIC/SOC

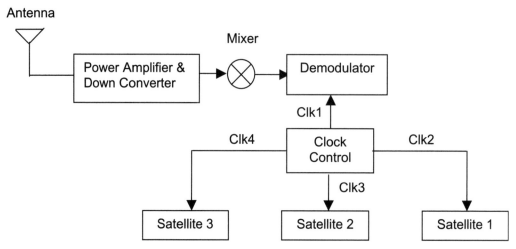

Fig. 5.15 Multiple Clock Domains in a GPS

shown success in dealing with complex designs having multiple clock domains. It is important to note that clocks are not the only signals that can be gated to save power. Generally, signals with high frequency, such as address or data buses, are also good candidates for signal gating. Power saving becomes more significant when a group of signals share a control signal that has a lower frequency than the source signals.

Gate-Level Optimization

In order to minimize power in the gate level, load capacitances should be minimized by using less logic. Low-power EDA tools usually handle this as part of their optimization techniques. Let's look at a simple example to understand this concept. Here, gate minimization is achieved using proper Boolean functions, followed by appropriate use of don't cares in Karnaugh maps (K-maps).

As shown in Figure 5.16, the use of don't cares in a K-map can make a considerable change in gate count of a logic design. If don't cares are not used properly (as shown in Figure 5.16a), the number of gates used in a design can become very large, resulting in larger

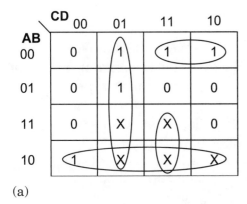 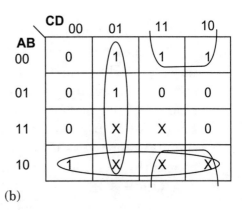

Fig. 5.16 Use of K-Maps in Minimizing Logic Gates in a Design
(a) Out = AB'+ C'D + A'B'C + ACD + AB'C (b) Out = AB' + C'D + B'C

silicon area and higher power consumption in designs. Figure 5.16b shows a more careful use of don't cares in K-maps, which illustrates a smaller number of logic gates resulting in less area and less power dissipation for the same design. Once the number of gates is minimized in a design, a low-power library has to be considered for various gates. Reduction of gate power is a relatively simple case of selecting particular libraries with the lowest power characteristics from among various choices of libraries provided by ASIC vendors.

Proper use of Boolean functions can also save power without reducing the number of logic gates. The following example illustrates the case of power saving without reducing the number of gates at logic level.

Consider the four inputs A, B, C, and D in Figure 5.17. Assuming input **A** has the highest activity compared to others, we can determine which of the following Boolean functions represents a lower power implementation (if all gates dissipate the same amount of power per signal transition).

1. Out = AB + AC + CD
2. Out = A (B + C) + CD
3. Out = AB + C (A+ D)

Figure 5.17 provides the logic gates required to build each of the above implementations.

As shown in Figure 5.17, all three implementations require four logic gates. However, the number of logic gates that signal **A** (i.e., the signal with the highest activity) propagates through represents the amount of power dissipation by each implementation. Signal **A** in Figure 5.17a propagates through three logic gates. In Figure 5.17b, signal **A** propagates through two gates and, finally, in Figure 5.17c, it passes through four logic gates. Therefore, the second implementation, **Out = A (B + C) + CD**, is the most efficient one for low power.

Another gate-level optimization approach is to reduce the transistor sizes. This results in decreasing input capacitances that could as well be the load capacitances for other gates. Although reducing the transistor sizes reduces the capacitance, it also reduces the current derive of each transistor and makes circuits operate slower. For that reason, transistor resizing is an appropriate technique only for noncritical paths in a design.

Power Estimation

To optimize power properly, circuit designers should be able to predict the power at early stages of their design. Power estimation can provide a design criterion for designers by evaluating various design options and the related power efficiencies before they commit to a specific low-power design methodology. Therefore, estimating the power consumption and analyzing the effect of various modifications at different levels of design abstraction are crucial to a successful low-power design.

In order to estimate dynamic power in circuits, the switching activity of different nodes, β, from Equation 5.5 must be calculated. One of the common power-estimation techniques is simulation-based power estimation. In this technique, the circuit under test will be simulated with a set of input stimuli and its power consumption will be monitored continuously. The average power consumed using various input patterns provides an estimation for the power consumption of the device. This method can estimate power in a

5.3 LOW-POWER DESIGN TECHNIQUES AND METHODOLOGIES

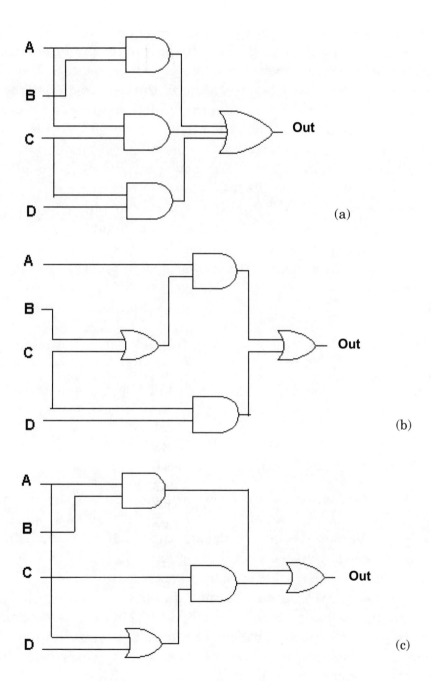

Fig. 5.17 Logic Required for Three Implementations of a Digital Circuit
(a) Out = AB + AC + CD (b) Out = A (B + C) + CD (c) Out = AB + C (A+ D)

variety of design styles; however, it is highly dependent on the pattern of input stimuli. This complexity of this technique increases with the number of input stimuli and size of the circuit. Simulation-based power estimation requires that all the input signals provide accurate results. The drawback is that individual blocks cannot be simulated without completing the rest of the design. This is a slow process, and providing all possible combinations of input stimuli for various node activities to estimate power accurately is a challenging task for low-power designers.

Power estimation followed by power optimization can be accomplished with special EDA tools that estimate and minimize the power dissipation at various levels of design abstraction. A more detailed discussion of low-power design tools is provided in the next section.

5.4 LOW-POWER DESIGN TOOLS

Numerous EDA tools are available to help IC designers achieve low-power designs. These tools are classified into two main categories:

- Power-analysis and power-estimation tools
- Power-optimization tools

Power-estimation tools estimate the power of a specific design by identifying its high power consuming modules at early stages of the design. These tools give IC designers the ability to make high-level design decisions to reduce power or leave the design untouched based on a set of specific power constraints.

Power-optimization tools come into play after the decision is made by IC designers to reduce the power. These tools automatically implement appropriate power-minimization techniques discussed in earlier sections of this chapter, and they provide optimal low-power designs based on the level of design abstraction that they target. The EDA tools for low power are categorized based on the abstraction layer in which they operate. These layers are as follows:

5.4 LOW-POWER DESIGN TOOLS

☞ Behavior/System
☞ RTL/Architecture
☞ Gate/Logic
☞ Transistor/Switch

Figure 5.18 shows how low-power EDA tools are used in a low-power design.

Table 5.4 provides a summary of some of the commercially available low-power EDA tools.

The two important issues to consider when evaluating the capabilities of low-power EDA tools are accuracy and speed of execution. Considering the role that each of these tools plays at differ-

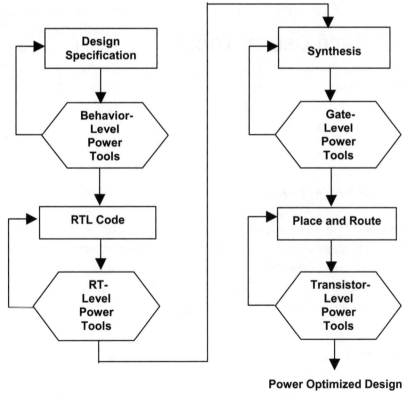

Fig. 5.18 Low-Power Design Flow Using EDA Tools

Table 5.4 Commercially Available Tools for Low-Power Design

EDA Tool	Abstraction Level	Function
DesignPower	RTL, Gate	Power Estimation
PowerCompiler	RTL, Gate	Power Optimization
PowerMill	Transistor	Power Estimation
PowerTheater Analyst	RTL, Gate	Power Analysis
PowerTheater Designer	RTL, Gate	Power Optimization
PrimePower	Gate	Power Analysis and Estimation

ent levels of design abstraction, there are always trade-offs between the accuracy and execution time. At the early stages of the design, IC designers need tools that are fast and capable of estimating or optimizing power easily for the entire design. Therefore, earlier in the design process and at high levels of design abstraction, the EDA tools need to be fast for analyzing the entire design. However, in later stages of design process, the tools need to be more accurate, and execution time becomes less crucial. Detailed information about some of these tools is available in Appendix A.

Behavior-Level Tools

This level of power estimation and optimization is the highest level of abstraction in design cycle, but is not supported by any commercial tools at present. Designers make spreadsheets based on various components, such as flip flops and memories in their design, and use power formulas to estimate the power.

RT-Level Tools

Register transfer, or architectural level, power estimation is a fast method that circuit designers can use to predict power consumption at the early stages of design hierarchy. RTL power-estimation tools can predict a circuit's power consumption using the number of transitions of different synthesis factors obtained from RTL simulations. This level of power estimation is less accurate compared to the other two levels defined earlier. However, it has two significant advantages that help designers to make design decisions at early stages of their design. These advantages are as follows:

- Faster runtime
- Power estimation early in the design cycle

The accuracy of this technique depends on the delay model used in the estimation tool.

Zero-delay model is the simplest model used in estimation tools. This model considers a value of zero for the delay of each gate. For simplification, the zero delay model disregards the glitching effects and calculates the circuit-node changes only once per clock cycle. This model does not provide realistic timing information about the gates in a design and underestimates the power in circuits.

Unit-delay model assumes a unique delay for all the gates in a circuit. Glitches in a design can be represented using this model. However, since the unit-delay model does not provide accurate timing for the gates, both underestimation and overestimation of power dissipation are anticipated.

Full-timing-delay model is the most accurate model for power estimation. Glitches can be perfectly represented in designs using full-timing-delay models. Therefore, no overestimation or underestimation of dissipated power is will occur.

Most RTL optimization tools automatically implement clock gating to synchronous registers in order to save power. Throughout the power-optimization period, the power information is updated continuously until the optimal low-power solution is achieved.

Examples of low-power estimation tools are DesignPower from Synopsys and PowerTheater Analyst from Sequence Design, Inc. The popular power-optimization tools are PowerCompiler from Synopsys and PowerTheater Designer from Sequence Design, Inc.

Gate-Level Tools

Gate- or logic-level power-estimation tools are less accurate compared to transistor-level tools but have considerable enhancement in runtime. These tools estimate power consumption due to the transitions made by every node in gate-level simulation. Gate-level power-estimation tools operate on a gate-level netlist, such as Verilog or VHDL, and calculate the power consumed at each node using Equation 5.5. The sum of all node powers represents the estimated power of circuit as shown in Equation 5.7.

$$P_{est} = 1/2 \sum \beta \cdot C \cdot V_{DD}^2 \cdot F \qquad \text{(Eq. 5.7)}$$

This level of power estimation is more accurate than RTL estimation methods and is used in later stages of design hierarchy when the design is completed, synthesized, and simulated. Examples of these tools include PowerTheater Analyst and PrimePower.

Transistor-Level Tools

Transistor or switch-level power-estimation tools are very accurate. These tools can model each circuit element precisely and estimate power consumption within a few percentage points of precision. However, long runtime restricts their use to small designs. Transistor-level power-optimization tools automatically resize the transistors in a design to meet power requirements estimated by transistor-level power-estimation tools. They compute the slack time of each gate. For negative slacks, these tools upsize the transistors, and for positive slacks, they downsize each transistor. Also, signals with high-transition level are automatically assigned with short wires.

Although these tools are very accurate, they are not very popular among ASIC and SOC designers for the following two reasons. One is the long runtime that restricts their use to small circuits. The other is that most ASIC vendors do not provide transistor level netlists.

Examples of these tools include HSPICE and PowerMill from Synopsys Inc.

5.5 Tips and Guidelines for Low-Power Design

The goal of this section is to summarize the tips and guidelines for low-power design of ASICs and SOCs.

- ☞ The most effective power-optimization techniques are the higher level ones. These are algorithmic and architectural optimization techniques.
- ☞ Use low-power process and libraries. There are low-power standard cell libraries for 0.18 μm such as Xemics CooLib. (Refer to reference 16 for more information.) The low-power libraries should be used in conjunction with a low-power process that is available from most ASIC vendors.
- ☞ Decrease the dynamic power by reducing all of the terms in the fundamental equation of power:

$$P = 1/2 \cdot \beta \cdot C \cdot V_{DD}^2 \cdot F$$

- ☞ Apply the following for your supply voltages:

  Lower the supply voltage for the entire chip when possible.

  Use low-supply voltage for noncritical paths and high-supply voltage for components on critical paths.

- ☞ Gate your clocks and high-activity signals whenever possible to prevent excessive switching activity in synchronous designs.
- ☞ Reduce the power consumed in memory blocks/cores using the following approaches:

  Minimize the number of accesses to main memory by having more code in cache.

  Split large memories to smaller modules; for example, a 256k × 32 RAM can be split into two 128k × 32 RAMs.

Segment memories with individual clocks and put each segment in sleep mode when idle.

☞ Minimize the number of transitions in high-activity bus interconnects using the following methods:

Partition a wide bus to multiple narrow buses since the bus width is directly related to power dissipation.

Use bus-encoding techniques such as gray coding, bus inverting, and partial bus inverting.

☞ Use power management techniques to shut down or minimize the switching activities of certain blocks/cores in your SOC (i.e., put them in sleep mode). This is more effective through software solutions.

☞ EDA power estimation and power-optimization tools should be used in all design abstraction levels. However, these tools are more effective early on at the algorithm or architecture phases of a design.

5.6 SUMMARY

In this chapter, sources of power dissipation in CMOS devices were discussed.

The fundamental power equation

$$P_{Dynamic} = 1/2 \cdot \beta \cdot C \cdot V_{DD}^2 \cdot F$$

is the basis for all power reduction techniques. Each term in the equation can contribute to power savings.

Several methods of power optimization at different levels of design abstraction for ASICs and SOCs were explained. These techniques are:

☞ Algorithm-level optimization
☞ Architecture-level optimization

- RT-level optimization
- Gate-level optimization

In addition to power-optimization techniques, the simulation-based power-estimation method was discussed. Low-power EDA tools for predicting power at different levels of design abstraction were covered. Detailed information about some of the popular low-power design (estimation and optimization) tools is available in Appendix A.

Tips and guidelines for low-power designs were also covered.

5.7 REFERENCES

1. J. M. Rabaey and M. Pedram. *Low-Power Design Methodologies.* Norwell, MA: Kluwer Academic Publishers, 1996.
2. K. Roy and S. C. Prasad. *Low-Power CMOS VLSI Circuit Design.* New York: Wiley Interscience Publication, 2000.
3. C. Settles and S. Emerson. "Centralized Clock Control Tames Power Consumption in SOCs." *Portable Design Magazine*, pp. 60–65, January 2000.
4. C. Piguet. *Low-Power DSP Processors.* DSP Course, March 1999, Monterey, CA.
5. R. Burch, F. N. Najm, P. Yang, and T. Trick. "A Monte Carlo Approach for Power Estimation." *IEEE Transactions on VLSI Systems*, pp. 63–71, March 1993.
6. T. L. Chou and K. Roy. "Accurate Power Estimation of CMOS Sequential Circuits." *IEEE Transactions on VLSI Systems*, pp. 369–380, September 1996.
7. F. N. Najm. "A Survey of Power Estimation Techniques in VLSI Circuits." *IEEE Transactions on VLSI Systems*, pp. 446–455, December 1994.
8. M. J. Irwin. "Low-Power Design for Systems on a Chip." Department of CSE, Penn State University. *www.cse.psu.edu/~mji*
9. Synopsys' PowerCompiler. *www.synopsys.com/news/press/pc_pr.html*
10. J. Frenkil. "Low-Power Design Tools, Where Is the Impact?" Design Automation Conference, June 1997.

11. M. Poncino. "Power Optimization of Memory Processor Interfaces in Embedded Systems." *www.eda.sci.univr.it*
12. M. A. Cirit. "Estimating Dynamic Power Consumption of CMOS Circuits." *IEEE International Conferences on Computer Aided Design*, pp. 534–537, November 9–12, 1987.
13. R. Burch, F. Najm, P. Yang, and D. Hocevar. "Pattern-Independent Current Estimation for Reliability Analysis of CMOS Circuits." *25th ACM/IEEE Design Automation Conference,* Anaheim, CA, pp. 294–299, June 12–15, 1988.
14. F. Najm, R. Burch, P. Yang, and I. Hajj. "CREST—A Current Estimator for CMOS Circuits." *IEEE International Conference on Computer Aided Design*, Santa Clara, CA, pp. 204–207, November 7–10, 1988.
15. "Managing Power in Ultra Deep Submicron ASIC/IC Design." *www.synopsys.com/products/power/low_power_wp.pdf*
16. *www.xemics.com*
17. K. Itoh, K. Sasaki, and Y. Nacagome. *IEEE Symposium of Low Power Electronics,* pp. 84–85, October 1994.
18. "Power Analysis Technology Backgrounder," Synopsys Inc. Mountain View, CA, September 1995.
19. C. Deng. "Power Analysis for CMOS/BiCMOS Circuits." In *Proceedings of the 1994 International Workshop on Low-Power Design*, pp. 3–8, April 1994.
20. K. Itoh, K. Sasaki, and Y. Nakagome. "Trends in Low-Power RAM Circuit Technologies." *Proceedings IEEE,* April 1995.
21. H. W. Johnson and M. Graham. *High-Speed Digital Design.* Upper Saddle River, NJ: Prentice-Hall, 1993.
22. A. Chandrakasan, R. Allmon, A. Stratakos, and R. Brodersen. "Design of Portable Systems." *Proceedings CICC Conference*, May 1994.
23. A. Chandrakasan, S. Sheng, and R. Brodersen. "Low-Power CMOS Digital Design." *IEEE Journal of Solid State Circuit*, SC-27, pp. 658–691, 1992.
24. W. C. Athas. *Low-Power Design Methodologies.* Norwell, MA: Kluwer Academic Press, 1995.
25. N. H. E. Weste and K. Eshraghian. *Principles of CMOS VLSI Design: A System Perspective.* Reading, MA: Addison-Wesley Publishing Company, 1994.
26. M. Takada. *Low-Power Memory Design.* IEDM Short Course Program, 1993.

27. F. Ferguson and A. Mauskar. "High Level Design for Low Power." *Proceedings of Electronics Design Automation and Test Conference*, August 1995.
28. D. K. Liu and C. Svensson. "Power-Consumption Estimation in CMOS VLSI Chips." *IEEE J. Solid State Circuits*, vol. 29, pp. 663–670, 1994.
29. S. R. Powell and P. M. Chau. "Estimating Power Dissipation of VLSI Signal Processing Chips: The PFA Technique." *VLSI Process.*, vol. 7, pp. 250–259, 1990.
30. J. Lipman. "EDA Tools Let You Track and Control CMOS Power dissipation." *EDN Magazine*, November 1995.
31. M. Farrahi, G.E. Tellez, and M. Sarrafzadeh. "Memory Segmentation to Exploit Sleep Mode Operation." *Proceedings of the Design Automation Conference*, June 1995.

APPENDIX A

Low-Power Design Tools

PowerTheater[1]

PowerTheater is a comprehensive set of power tools that create maximum power efficiency for SOC designs at the architectural level, RT level, and gate level. This family of full-chip power tools can be used throughout the IC design process. PowerTheater products (PowerTheater Analyst and PowerTheater Designer) analyze, display, and help the user to reduce the power for the whole chip and each individual module. These tools interface directly with Verilog and VHDL simulators to capture behavioral and gate-level simulation activity.

Key Features

- SOC RTL power analysis
- Flexible and easy-to-use RTL power optimization
- Handles clock, memory, data-path control logic, and I/O
- De facto industry standard for RTL power design
- Versatile graphical analysis environment
- Accurate gate-level power verification

---

1. Courtesy of Sequence Design, Inc. Portions reprinted with permission.

Specifications

Platforms

- Sun Solaris
- HP UX

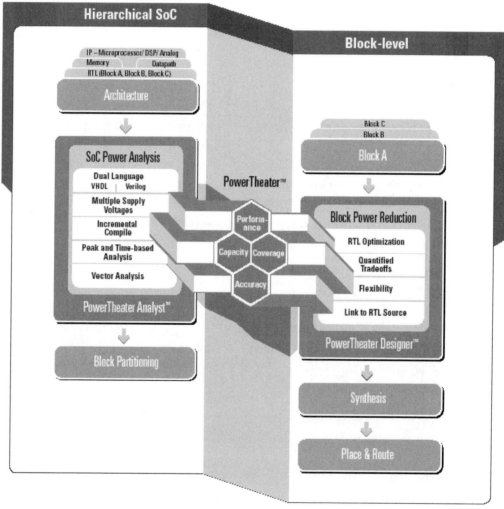

Fig. A.1 PowerTheater Block Diagram

Memory Configuration

- 512 Megabytes minimum
- 1 Gigabyte recommended

Input Formats

- Verilog
- VHDL
- ALF vendor libraries
- Synopsys Liberty (.LIB) vendor libraries
- OLA vendor libraries

Output Files

- ASCII report files

Integration

- PLI and VPI for Verilog simulators
- FLI interface for VHDL
- VCD supported

POWERTHEATER ANALYST

PowerTheater Analyst, built with proven Watt Watcher technology, represents a superior alternative to tedious and error-prone manual methods, such as spreadsheet analysis, as well as gate- and transistor-level methods which require the design to be synthesized in pieces and then simulated at very detailed levels of abstraction. Using this capability at both RT and gate level, designers can perform detailed power analysis for the entire chip or any set of sub-blocks, including memory, I/O, logic, and clock trees. Peak and time-based power is reported on a user-defined time interval, down to the Verilog or VHDL simulation resolution. PowerTheater Analyst

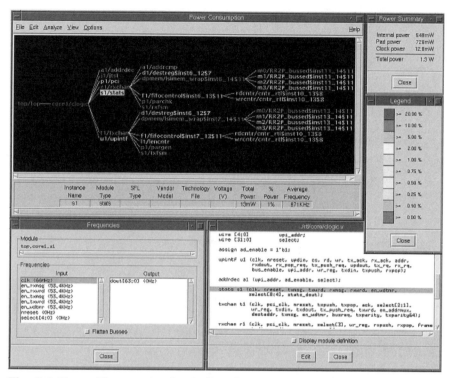

Fig. A.2 PowerTheater Analyst's Graphical User Interface

addresses issues such as power bus sizing, electromigration, and reliability at the RT level, before the design is synthesized.

Features and Benefits

- ☞ Accurate RTL power estimation helps designers minimize power early in the design cycle.
- ☞ Versatile graphical analysis environment lets designers assess trade-off options quickly and intuitively.
- ☞ Peak and time-based power analysis allows designers to pinpoint power problems and understand the details of power consumption.

- Vector analysis capability to determine vector coverage and testbench quality.
- Fast, high-capacity RTL, gate, and mixed full-chip analysis for multimillion-gate designs covers all major contributors of power dissipation including:

    Clocks, including built-in clock-tree estimation

    Memory, including single and multiport SRAM and DRAM

    Data path and control logic

    I/O, including multiple voltage

- Supports Verilog, VHDL, and mixed-language analysis.
- Incremental compile and split activity file processing increase analysis performance and capacity.
- Comprehensive detailed reports in formatted text and HTML.

POWERTHEATER DESIGNER

The PowerTheater Designer product uses Watt Watcher's proven RTL estimation technology to build a detailed, quantitative map of the power in the design. PowerTheater Designer then invokes a suite of patent-pending agents called WattBots that automatically measure the impact of many potential power-saving architectural changes. Each WattBot is designed to identify a specific type of power reduction opportunity. The suite of WattBots covers all major types of circuitry in the design, including control, datapath, I/O, memory, and clocks. For each opportunity identified, PowerTheater Designer proposes specific RTL design modifications and quantifies the power savings that would result, together with certain potential trade-offs, such as any area penalty. Users then choose which changes to implement and which to reject, based on their power targets and design flow needs.

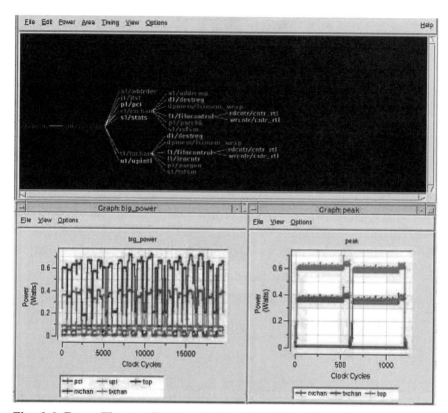

Fig. A.3 PowerTheater Designer's Graphical User Interface

PowerTheater Designer operates at the RT level and locates much larger power reductions than are available via gate-level techniques such as local clock gating and buffer resizing.

Features and Benefits

- ☞ RTL power reduction for system-on-a-chip integrated circuits
- ☞ Allows presynthesis optimization covering all major contributors of power dissipation:

  memory

  clocks

> data paths
>
> control logic
>
> I/O

- 20–40% power reduction on large ICs, vs. 5–10% for gate/synthesis-based tools
- Supports both Verilog and VHDL languages
- Versatile graphical analysis environment lets designers assess trade-off options quickly

APPENDIX B

Open Core Protocol (OCP)[1]

The Open Core Protocol (OCP) delivers the first openly licensed, core-centric protocol that comprehensively fulfills system-level integration requirements. The OCP defines a comprehensive, bus-independent, high-performance, and configurable interface between IP cores and on-chip communication subsystems. A designer selects only those signals and features from the palette of OCP configurations needed to fulfill all of an IP core's unique data, control, and test-signaling requirements. Existing IP cores may be inexpensively adapted. Defining a core interface using the OCP provides a complete description for system integration, and enables core reuse and test reuse without rework. The OCP supports very high transfer performance, with data-transfer models ranging from simple request-grant through pipelined request-response to more complex out-of-order operations. OCP provides a clear delineation of design-responsibility boundaries between core authors and SOC integrators.

---

1. Courtesy of OCP International Partnership Association, Inc (OCP-IP). Portions reprinted with permission.

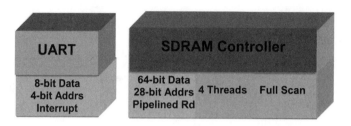

Very simple through highly complex core signaling can all be captured by this single protocol

HIGHLIGHTS

The OCP promotes IP core reusability and reduces design time, design risk, and manufacturing costs for SOC designs. The OCP is focused exclusively on the needs of an IP core; nothing about the OCP is bus or application specific.

- Enables IP core creation to be independent of system architecture and application design
- Describes all intercore communications
- Optimizes die area by configuring into the OCP only those features needed by the core
- Uses timing guidelines to assure IP core interoperability:

    Level 2—highest performance interface timing

    Level 1—conservative timing for effortless core connect

    Level 0—protocol without timing specified (especially useful for verification and simulation tools)

CAPABILITIES

The OCP captures all core characteristics without restricting system arbitration, address map, etc.

- Small set of mandatory signals, with a wide range of optional signals.
- Core-specific data and address widths.
- Structured method for inclusion of sideband signals: high-level flow control, interrupts, power control, device-configuration registers, test modes, etc.
- Synchronous unidirectional signaling allows simplified implementation, integration, and timing analysis.
- Transfers may be pipelined for reduced latency.
- Optional burst transfers for higher efficiency.
- Multiple concurrent transfer sequences may be managed with thread identifiers, for out-of-order completion.
- A connection identifier may be used to provide end-to-end identification for targets desiring to distinguish initiators for service prioritization, etc.
- OCP is a functional superset of the VSIA's Virtual Component Interface (VCI), adding configurable sideband control signaling and test harness signals.

ADVANTAGES

- Eliminates the ongoing task of (re)defining interface protocols, then verifying, documenting, supporting them.
- OCP readily adapts to support new core capabilities.
- Testbench portability simplifies reverification.
- Limits test suite modifications for core enhancements.
- Any bus structure can be interfaced to the OCP.
- Delivers industry standard flexibility and reuse.
- Symmetrical signaling enables direct point-to-point communication between two cores (without on-chip bus).

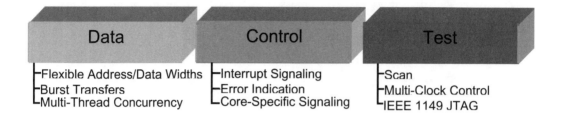

The full Open Core Protocol specification is available at: *www.ocpip.org*

KEY FEATURES

Basic OCP

- Master-slave interface with unidirectional signals
- Driven and sampled by the rising edge of the OCP clock
- Fully synchronous, no multicycle timing paths
- All signals are strictly point-to-point (except clock and reset)
- Simple request/acknowledge protocol

    Supports data transfer on every clock cycle

    Allows master or slave to control transfer rate

- Configurable data word width
- Configurable address width
- Pipelined or blocking reads
- Specific description formats for core characteristics, interfaces (signals, timing, and configuration), performance

Simple Extensions Enhance Performance

- Burst codes link related transfers into complete transaction

- Burst transactions supported:

    Sequential (defined or undefined length)
    Streaming (FIFO)
    Core-specific (cache lines)

- Pipelined (address ahead of data) writes
- Aligned or random byte enables
- Read dataflow control
- Address space definition

Complex Extensions—Enable Concurrency

- *Thread identifiers* enable:

    Interleaved burst transactions
    Out-of-order transaction completion

- *Thread busy* notification prevents interface blocking
- *Connection identifiers* enable:

    End-to-end system initiator identification
    Service priority management by system targets

Sideband Extensions—Dedicated Signaling

- Core-specific, user-defined signals:

    System event signals (e.g., interrupts error notification)
    Synchronous reset
    Data transfer coordination (e.g., high-level flow control)

Debug and Test Interface Extensions

- Support structured full or partial scan test environments

- *Scan* pertains to internal scan techniques for a predesigned hard core or end user-inserted into a soft core.
- *Clock Controls* are used for scan testing and debug, including multiple clock domains
- *IEEE 1149* supports cores with a JTAG Test Access Port
- Configurable for JTAG and Enhanced JTAG-based debug for MIPS (EJTAG), ARM, TI DSP, SPARC and others

CoreCreator

OCP-IP offers its members the use of an EDA tool created by Sonics, Inc., named CoreCreator to automate the tasks of building, simulating, verifying, and packaging OCP-compatible cores. CoreCreator also supplies a protocol checker to ensure compliance with the OCP specification. IP-core products can be fully componentized by consolidating core models, timing parameters, synthesis scripts, verification suites, and test vectors.

APPENDIX C

Phase-Locked Loops (PLLs)

Most ASICs currently developed include one or more phase-locked loop (PLL) circuits. PLLs are used for a number of reasons including reduction of on-chip clock latency, synchronization of clocks between different ASICs, frequency synthesis, and clock-frequency multiplication. ASIC vendors offer various types of PLLs based on their frequency range, pin count, size, and stability.

PLL Basics

A simplified block diagram of a PLL is shown in Figure C.1. The reference clock, REFclk, is the input clock into the ASIC. The PLL tracks the reference clock and adjusts the phase of its output, PLL-out, such that REFclk and the feedback clock, FBclk, are in phase.

The phase detector compares the phase difference between the rising edge of REFclk and the rising edge of FBclk. When the two are not aligned, the phase-detector output changes to increase or decrease the voltage level on the output on the charge pump.

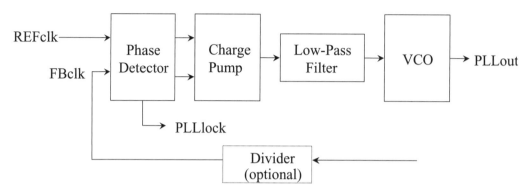

Fig. C.1 Phase-Locked Loop Block Diagram

The charge pump and low-pass filter convert the digital output of the phase detector into an analog voltage. The low-pass filter is used to control rate of change of input voltage into the voltage-controlled oscillator (VCO).

The VCO generates the clock output. The clock frequency changes as a function of the output of the charge pump.

If the PLL is being used as a frequency multiplier, the FBclk frequency is divided before being fed into the phase detector.

PLL IDEAL BEHAVIOR

A PLL will adjust the phase of its output such that its reference input REFclk and its feedback clock are perfectly aligned, or in phase.

Consider the ASIC PLL circuit shown in Figure C.2. In this ideal circuit, the PLL will perfectly align the arrival time of the feedback clock, t_{FB}, with the arrival time of the reference clock, t_{REF} ($t_{FB} = t_{REF}$). The output of the PLL is distributed throughout the ASIC with the use of a clock-distribution network that is perfectly balanced such that the delay from the PLL output to every ASIC register is equal ($dly_a = dly_b$).

The arrival time of the PLL output can be expressed as:

$$t_{PLLout} = t_{FB} - dly_b \qquad \text{(Eq. C.1)}$$

PLL Ideal Behavior

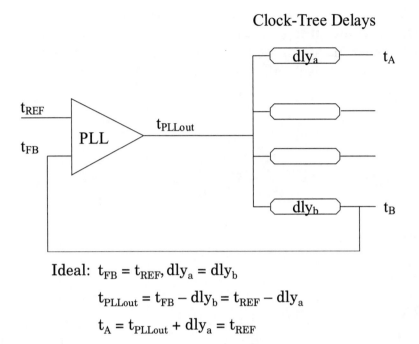

Fig. C.2 Ideal PLL Behavior

Since in this ideal case $t_{REF} = t_{FB}$ and $dly_a = dly_b$,

$$t_{PLLout} = t_{REF} - dly_a \qquad (Eq.\ C.2)$$

or

$$t_{PLLout} + dly_a = t_{REF} \qquad (Eq.\ C.3)$$

and the arrival time of the clock at a register, regA, can be expressed as

$$t_A = t_{PLLout} + dly_a = t_{REF} \qquad (Eq.\ C.4)$$

So, in this case, the arrival time of a clock at any register is perfectly aligned with t_{REF}. Note that the delay of the clock-distribution network is not a term of the arrival time, t_A. This shows how a PLL can be used to minimize the effects of on-chip clock latency.

Consider also two ASICs fed by the same clock, but whose clock-distribution networks have different latency times. Using PLLs in each of the ASICs will make the respective clocks appear to be in phase with each other.

$$\text{ASIC1}(t_A) = t_{REF} \qquad \text{(Eq. C.5)}$$

and

$$\text{ASIC2}(t_A) = t_{REF} \qquad \text{(Eq. C.6)}$$

hence,

$$\text{ASIC1}(t_A) = \text{ASIC2}(t_A) \qquad \text{(Eq. C.7)}$$

PLL Errors

PLLs don't exhibit ideal behavior. A PLL will not perfectly align t_{FB} and t_{REF}, and the delays through an ASIC clock-distribution network are never perfectly balanced. For purposes of timing analysis, the PLL errors to consider are as follows.

Static-Phase Error (SPE)

SPE is the fixed offset (error) between the rising edges of REFclk and FBclk caused by delay-path differences in the phase detector, process/voltage/temperature differences with the ASIC, and transition-time (slew-rate) differences on the REFclk and FBclk inputs to the PLL.

Long-Term Jitter (LTJ)

LTJ is a low-frequency drift in the offset between the rising edges of REFclk and FBclk.

Short-Term Jitter (STJ)

STJ is a high-frequency drift in the offset between the rising edges of REFclk and FBclk. This offset may vary from clock cycle to clock cycle. The STJ is a component of LTJ. For any given rising edge of PLLout, the next rising edge of PLLout will occur one clock period +/- STJ later.

Jitter is usually caused by power and ground noise introduced into the VCO.

These PLL errors are expressed as +/- delays. This means that because of nonideal behavior, t_{FB} may occur before or after t_{REF}.

Typical values for PLL errors are:

Static-Phase Error: ± 200 ps max

Long-Term Jitter: ± 250 ps max

Short-Term Jitter: ± 100 ps max

Consider the ASIC PLL circuit previously analyzed (see Figure C.3). In this case however, nonideal behavior is introduced such that the PLL exhibits SPE, LTJ, and STJ errors, and the ASIC clock-distribution network is not perfectly balanced (i.e., the clock skew is nonzero.) (Since STJ is a component of LTJ, STJ is not considered when analyzing the arrival time of a clock to any register, regA. The effects of STJ will be discussed later.)

The clock of regB is used as the feedback clock, FBclk, into the PLL. A delay exists between the arrival time of the clock at regB, t_B, and the arrival of the clock at the FB pin of the PLL, t_{FB}.

The difference between the arrival time of the clock at regA and the clock at regB represents the clock skew in the clock-distribution network.

The arrival time of the clock at regA can now be analyzed as follows:

$$t_{FB} = t_{REF} +/- \text{SPE} +/- \text{LTJ} \quad \text{(Eq. C.8)}$$

$$t_B = t_{FB} - \text{dlyFB} \quad \text{(Eq. C.9)}$$

$$t_A = t_B - \text{dly}_B + \text{dly}_A \quad \text{(Eq. C.10)}$$

Substituting and rearranging,

$$t_A = t_{REF} +/- \text{SPE} +/- \text{LTJ} - \text{dlyFB} + (\text{dly}_A - \text{dly}_B) \quad \text{(Eq. C.11)}$$

The greatest errors occur if the arrival time of the FBclk, t_B, is either the earliest or the latest point in the clock distribution. The arrival time, t_B, should be chosen as the midpoint in the clock-distribution network if the error between t_{REF} and t_A is to be minimized.

If intentional skewing of t_A relative to t_{REF} is desired, the feedback-path delay, dly_{FB}, can be adjusted to achieve the desired effect.

Consider again two ASICs with PLLs driven by the same clock REFclk with clock skews of ASIC1(skew) and ASIC2(skew). Because of the errors introduced by the PLLs and the clock-distribution networks on each ASIC, the arrival time of clocks with the ASIC may vary by as much as SPE + LTJ + ASIC1(skew) + ASIC2(skew). This can be a considerable amount affecting the ability to clock signals from one ASIC to the next, if the clock skews within each ASIC are not carefully controlled and if the FBclk is not carefully chosen.

SPE and LTJ affect the arrival time of a clock to a register relative to REFclk. This means that setup and hold times of ASIC I/Os are affected by SPE and LTJ. STJ effectively reduces the clock period, thereby requiring the ASIC to run at a higher frequency than required by REFclk. This must be taken into account when analyzing the critical path within the ASIC. For example, if the clock period of REFclk is cycREF, the ASIC must be analyzed using a clock period of (cycREF − STJ). If a multicycle path is analyzed using a period of n · cycREF, the use of a PLL requires that the same path run at n · (cycREF − STJ) or (n · cycREF − n · STJ).

PLL Errors

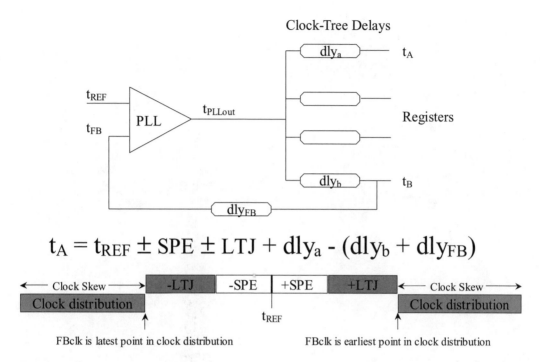

$$t_A = t_{REF} \pm SPE \pm LTJ + dly_a - (dly_b + dly_{FB})$$

Fig. C.3 Nonideal PLL Behavior

Glossary

10 BaseT: IEEE 802.3 specification for 10-Megabit Ethernet implementation on twisted-pair wiring.

100 BaseT: IEEE 802.3 specification for 100-Megabit Ethernet using level-5 UTP.

10G Ethernet: 10-Gigabit Ethernet supports the data rate of 10 Gbps and supports the features of the preceding Ethernet standards. The potential applications for 10-Gigabit Ethernet are Local Area Network (LAN), Metropolitan Area Network (MAN), and Wide Area Network (WAN).

AAL: The **A**TM **A**daptation **L**ayer is a service-dependent sublayer of DLL (Data Link Layer) that transfers data from various applications to the ATM layer using 48-byte ATM payload formats. ITU-T has recommended four types of AAL: AAL1, AAL2, AAL3/4, and AAL5.

AAL1: Layer 1 of AAL is used for connection-oriented, delay-sensitive applications with a constant data rate, such as uncompressed video transfer.

AAL2: Layer 2 of AAL is used for connection-oriented applications with variable data rate, such as voice transfer.

ADC: An **A**nalog-to-**D**igital **C**onverter converts analog signals to digital signals using sampling and quantization techniques.

ADPCM: Adaptive **D**ifferential **P**ulse **C**ode **M**odulation is a speech-compression technique that converts the analog voice signals to high-quality digital signals that can be transferred over 32-Kbps digital channels.

A-Law: An ITU-T standard for Europe to convert analog data to digital data using the Pulse Code Modulation (PCM) algorithm. North America and Japan use the µ-Law standard.

ASIC: An **A**pplication-**S**pecific **I**ntegrated **C**ircuit is a chip that is designed to satisfy a specific application's requirement.

ATM: Asynchronous **T**ransfer **M**ode is a broadband transmission system that is based on small, uniform packets and is widely used in LANs and WANs.

ATPG: Automatic **T**est **P**attern **G**eneration provides a set of test vectors to identify all the faults in a circuit.

BGA: Ball **G**rid **A**rray is a packaging methodology that reduces the area so more functions can be integrated on a single space. This packaging method provides higher performance due to the short distance that exists between the chips and solder balls. BGA chips are usually used on mobile applications since they are small and occupy less board area.

BIST: Built-**I**n **S**elf **T**est is a verification technique that allows a circuit, or portions of a circuit, to test itself and identify the faults. An output signal is sent when a fault is detected.

Boundary Scan: Boundary scan is a testing methodology that allows the boundary pins of a JTAG-compatible circuit to be tested using software control.

CELP: Code **E**xcited **L**inear **P**rediction is a voice-compression algorithm that is used for low bit-rate encoding/decoding. ITU standards for CELP are: G.728, G.729, and G.723.1.

Codec: **Co**der/**dec**oder is a compressor/decompressor device that converts specific types of analog data to digital data and vice versa. The analog data could be audio, speech, or video.

CPCI: A **C**ompact **P**eripheral **C**omponent **I**nterconnect or compact PCI is a high-performance interconnect bus based on PCI standards, which can support 8 PCI slots on a single bus.

DAC: A **D**igital-to-**A**nalog **C**onverter is a device that converts digital information to analog voltage levels. One example is converting digital information from a CD to analog audio signals.

DFT: **D**esign **F**or **T**est is a chip-design technique that incorporates testing into a design from the very beginning of the design process to reduce the test-generation complexity at later stages.

DIP: **D**ual **I**n-line **P**ackage is a packaging method that uses two parallel rows of pins. Most DIP devices have 14 to 16 pins.

DLL: **D**ata **L**ink **L**ayer is the second layer in OSI model and defines how data moves to or from physical media to upper layers. It also provides error detection and flow control. DLL contains two sublayers: Media Access Control (MAC) and Logical Link Control (LLC).

DMA: **D**irect **M**emory **A**ccess is a data-transfer method that allows information to be transferred between several memories or other peripheral devices without the need to go through the CPU each time. DMA-based devices have faster data transfer compared to devices that require data to pass through the main CPU.

DRC: **D**esign **R**ule **C**heck is a process of checking the final semiconductor layout against a set of physical design rules. These rules ensure that the design will not fail due to short circuits or process faults.

DSP: A **D**igital **S**ignal **P**rocessor is a specialized processor used for analyzing digital signal-processing algorithms such as voice processing and compression.

DTMF: **D**ual-**T**one **M**ulti **F**requency is a method used for voice processing and for use of two simultaneous tones such as in touchtone.

DUT: **D**evice **U**nder **T**est is the integrated circuit or part of the circuit that is of interest to the designer for testing.

EDA: **E**lectronic **D**esign **A**utomation tools are special software tools that are used to design and verify integrated circuits and systems.

EDIF: **E**lectronic **D**esign **I**nterchange **F**ormat is an industry standard format for transferring and interchanging design data. EDIF files can be created from a design schematic, VHDL, or Verilog code that has been processed with through synthesis tools.

ERC: **E**lectrical **R**ule **C**hecking is the process whereby circuit designers check the electrical rules provided by ASIC vendors. Examples of ERC violation are open input, short circuit, NMOS connected to Vdd, and PMOS connected to Gnd.

Ethernet: Ethernet is a LAN transmission standard that uses the Collision Sense Multiple Access with Collision Detection (CSMA/CD) method. Ethernet is similar to IEEE 802.3 standard.

FIFO: **F**irst-**I**n **F**irst-**O**ut is a buffer where data is processed in the order they were received. It is the opposite of a Last-In First-Out (LIFO) data structure.

FSM: A **F**inite **S**tate **M**achine is a useful method in digital design and is a function that maps a set of input events to their matching output events.

G.711: An ITU Pulse Code Modulation (PCM) standard for voice encoding/decoding at 48 to 64Kbps. It is a specification for A-Law/μ-Law coding.

G.723.1: An ITU standard for compression/decompression of speech and audio signals with very low bit rates such as 5.3 and 6.3Kbps. The lower bit rate is CELP and provides more flexibility for designers while the higher bit rate has better quality and is based on Multipulse Maximum Likelihood Quantization (ML-MLQ) technology.

G.726: An ITU standard for ADPCM voice compression from a 64-Kbps A-law /μ-law channel to 40-, 32-, 24-, or 16-kbps channels.

G.728: An ITU standard for low-delay CELP voice compression.

G.729: An ITU standard for CELP voice compression at 8Kbps.

Gateway: A point at a network that provides an entrance to another network and transfers packets from one network to another on Internet. Gateway is used in applications that route information from one network to another like packet switched networks.

GDSII: GDSII is a binary file format that is used to transfer the layout circuit-design information.

H.100: H.100 is a telephony bus, such as ribbon cable bus, and is used to transfer voice over a Compact Peripheral Component Interconnect (CPCI).

H.110: H.110 is derived from H.100 and uses a Time-Division-Multiplexing (TDM) bus for telephony applications such as VoIP.

HLB: Hierarchical **L**ogic **B**locks are used in hierarchical design methodology and are blocks that can be independently laid out as hard macros.

IEEE802.3: The IEEE LAN standard which has different physical-layer specifications that define Ethernet, for example 10Base2, 10Base5, 10BaseF, 10BaseT, and 10Broad36. Various specifications, such as 100BaseT and 100BaseT4, are available for fast Ethernet.

ILM: An **I**nterface **L**ogic **M**odel improves chip-level timing-analysis performance mainly by reducing the size of the netlist. ILMs are used for static timing analysis with Synopsys Primetime. These have replaced the traditional STAMP models.

IP: Intellectual **P**roperty blocks are predesigned and verified blocks of logic that can be reused for multiple designs.

IP: Internet **P**rotocol provides features such as addressing, security, and type of service specification.

ITU: International **T**elecommunication **U**nion is the organization that sets the international telecommunication standards.

ITU-T: A sector in ITU, it is responsible for standardization of worldwide telecommunications.

JTAG: The **J**oint **T**est **A**ction **G**roup is an IEEE standard that controls the pins of compliant devices on a Printed Circuit Board (PCB) to ensure the board-level continuity.

LVS: **L**ayout **V**ersus **S**chematic is a method that compares the layout netlist to a schematic netlist to ensure that the layout matches the schematic.

MAC: **M**edia **A**ccess **C**ontrol is an IEEE specification for the lower half of DLL that defines the access-control protocols in an OSI model.

MicroNetwork: A MicroNetwork is a heterogeneous integrated network that unifies, decouples, and manages all of the communication between processors, memories, and input/output devices on an SOC.

MII: A **M**edia **I**ndependent **I**nterface is a standard in Ethernet devices that interconnects the MAC sublayer and physical layer (PHY) despite the difference in media.

MPEG: The **M**oving **P**icture **E**xperts **G**roup is a video-compression standard intended to reduce the storage requirements for full-motion video. MPEG includes several standards such as MPEG-1, MPEG-2, and MPEG-4. MPEG-1 provides CD-ROM-quality storage of video; MPEG-2 is used for Set-Top Boxes (STB), DVDs, and HDTVs; and MPEG-4 is used for seamless transfer of audio/video information over Internet and wireless channels.

μ-Law: An ITU-T standard for North America and Japan to convert analog data to digital data using PCM algorithm. Europe follows the A-Law standard.

MVIP: **M**ulti **V**endor **I**ntegration **P**rotocol is a subset of the H.100 bus standard and is used for transferring data between switching and telephony processing boards on a PC.

OCB: **O**n-**C**hip **B**uses are used on complex ASICs and SOCs. System buses such as ARM AHB or MIPS system bus and peripheral buses such as PCI bus or ARM APB bus are examples of OCBs.

OCP: **O**pen **C**ore **P**rotocol is a core-centric protocol that is a bus-independent, high-performance, and configurable interconnect between various IP cores and on-chip communication subsystems. OCP is a functional superset of the VSI Alliance virtual-component-interface (VCI) specification (see Appendix B).

OIF: **O**ptical **I**nternetworking **F**orum is a worldwide nonprofit organization that promotes development of products and services for optical internetworks.

OSI: **O**pen **S**ystems **I**nterconnection defines a reference model on how data should be transferred between two points in telecommunication networks. It contains seven layers: Physical, Data Link, Network, Transport, Session, Presentation, and Application.

Packet: Packets are small pieces of data of a fixed size that transfer data over networks. Packets usually contain header information and payload data. The header information provides the information on origin, destination, and synchronization. The payload provides the data.

PCI: A **P**eripheral **C**omponent **I**nterconnect is a local bus standard designed by Intel for PCs that provides high-speed connection between PCs and several peripheral devices.

PCM: **P**ulse **C**ode **M**odulation is a method for converting and transmitting analog signals commonly used by telephone companies. In this method analog signals are sampled at specific intervals to generate pulses, which are coded to represent the original analog signal.

PGA: A **P**in **G**rid **A**rray is a type of integrated circuit socket used for chips that have many pins since the connecting pins are at the bottom of the chip in squares with separation of only 0.1 inch in each direction.

PLL: **P**hase **L**ocked **L**oops are used for reduction of on-chip clock latency, synchronization of clocks between different ASICs, frequency synthesis, and clock-frequency multiplications.

PSTN: A **P**ublic **S**witched **T**elephone **N**etwork is a worldwide voice telephone network.

QFP: Quad **F**lat **P**ack is a surface-mount-technology (SMT) package for chips and is rectangular or squared with lead on all four sides.

RMII: Reduced **M**edia-**I**ndependent **I**nterface is used in 10M and 100M Ethernet which offers faster transmission to MII with lower pin count.

SCSA: A **S**ignal **C**omputing **S**ystem **A**rchitecture transmits information on a computer telephony system for multiple client applications.

SDF: A **S**tandard **D**elay **F**ormat is a standard format in the electronic industry for defining place and route delays in a design.

SFI-4: Serdes-to-**F**ramer **I**nterface Level **4** is an OIF standard that is optimized for pure data transfer and describes the data transfer with clock rates at the actual line rates.

SFI-5: Serdes-to-**F**ramer **I**nterface Level **5** is an OIF standard for 40-Gbps packet and cell transfer in applications such as Packet-over SONET/SDH.

SiliconBackplane: SiliconBackplane is an example of a MicroNetwork which is licensed by Sonics, Inc. (see Section 1.3).

SOC: A **S**ystem **O**n a **C**hip is a system on an IC that integrates software and hardware intellectual property using more than one design methodology for the purpose of defining the functionality and behavior of the proposed system.

SOP: A **S**mall **O**utline **P**ackage is a type of packaging that has two rows of pins closely spaced with each other.

SPI-4P2: System **P**acket **I**nterface Level **4 P**hase **2** is a 10 Gbps electrical interface between the physical and data-link layers for SONET/SDH systems with independent transmit and receive interface.

STA: **S**tatic **T**iming **A**nalysis is a static verification method that verifies the delays within a device. It is capable of verifying every path and can detect serious problems such as glitches on the clock, violated setup and hole times, slow paths, and excessive clock skew.

STAMP: STAMP models are static timing models for complex blocks, such as DSPs and RAMs. STAMP models are created by core or technology vendors who provide database (.db) files for their customers as their timing models.

STB: A **S**et **T**op **B**ox is used to receive and decode the digital TV signals from cable or satellite for digital home entertainment systems.

TAP: A **T**est **A**ccess **P**ort consists of four pins defined by the IEEE1149.1 standard and provides boundary scan and other test interfaces for a circuit. These pins are TCK, TMS, TDI, and TDO.

TAT: **T**urn **A**round **T**ime. This term is frequently used in the semiconductor industry for the time it takes semiconductor vendors to make an ASIC prototype and a working part.

TCP: **T**ransfer **C**ontrol **P**rotocol is a transfer-layer protocol that provides retransmission sequencing for a reliable, connection-oriented transmission between two networks.

TCP/IP: **T**ransmission **C**ontrol **P**rotocol/**I**nternet **P**rotocol is a set of standards for network communications between multiple applications such as computers connected to networks.

TDM: **T**ime **D**ivision **M**ultiplexing is a technology that transmits multiple channels of information, such as voice, video, and data, over a single transmission path.

TSI: **T**ransmitting **S**ubscriber **I**dentification is a signal that shows the identification of the transmitting terminal.

UDP: **U**ser **D**atagram **P**rotocol is a network protocol for connectionless and unreliable transmissions that is used for exchange of replies between networks.

USB: A **U**niversal **S**erial **B**us is a plug-and-play external interface between a computer and external peripherals using a bidirectional cable.

UTOPIA: The **U**niversal **T**est and **O**peration **PHY** **I**nterface for **A**TM is an electrical interface for transmission of information on devices connecting to an ATM network.

VCI: The **V**irtual **C**omponent **I**nterface is a standard for bus architecture defined by VSIA for intellectual property interactions.

Vocoder: A **Vo**ice **Coder** is a speech/voice compressor/decompressor system that converts analog speech to digital signals and vice versa.

VoIP: **V**oice **O**ver **I**nternet **P**rotocol is a technology that is used to transmit voice over digital networks on Internet with high quality and low cost.

VoN: **V**oice **O**ver **N**etwork is a method that uses packetized voice data for transmission over a network in Internet telephony technology.

VSIA: **V**irtual **S**ocket **I**nterface **A**lliance is an organization that promotes standards to design SOCs with reusable intellectual properties.

WAN: **W**ide **A**rea **N**etwork is a system that connects LANs together over a long-distance medium.

WLM: A **W**ire **L**oad **M**odel relates a net's estimated length to estimated capacitance and resistance in a synthesis tool to provide an approximation of wire delays.

XNF: **X**ilinx **N**etlist **F**ormat was developed by Xilinx as a hardware-description language. XNF can be converted to other hardware-description languages such as Verilog.

Index

A

AC characteristics, 21
Amoeba technology, 99
Analog-digital converter (ADC), 23
Antenna Check, 87
ASIC, 1, 21, 22
Automatic Test Pattern Generation (ATPG), 27, 37
Automation, 54

B

Back end, 18, 81, 83, 93, 95
Ball Grid Array (BGA), 22
Binary coding, 126
Block-based design, 105–106
Boolean function, 137
Boundary scan, 37–38
Buffer-tree network, 103
Built-In Self Test (BIST), 15, 27, 37
Bus-interconnect, 117
Bus-invert (BI) coding, 127

C

CAD tools, 84
Clock design, 90
Clock-gating, 27
Clock-tree synthesis, 103
CMOS power dissipation, 114
Code coverage, 51
Code excited linear prediction (CELP), 5, 121
Control flow analysis, 48

Crosstalk, 85, 86
CVS, 54

D

Data flow analysis, 48, 61
DC characteristics, 21
Deep submicron (DSM), 95
Delay calculation, 18
Design analysis, 32
Design flow, 10, 25
Design for integration, 44
Design for test (DFT), 15, 27, 37–40, 53
Design methodology, 16
Design partitioning, 100
Design Rule Check (DRC), 86
Design rule constraints, 26
Design validation, 34
Detail routing, 84
Device Under Test (DUT), 39
DFT sign-off, 87
Digital signal processing (DSP), 5, 73, 121
Digital-analog converter (DAC), 23
Dual in-line, 22
Dynamic power dissipation, 113, 116

E

EDA tools, 84
EDIF, 32
Electromigration, 86
Electrical Rule Check, 87
Embedded array, 32
Ethernet, 4
External interface emulation, 48

F

FIFO, 59, 78
Finite-impulse-response (FIR) filter, 121
Finite-state-machine (FSM) encoding, 124
Flip flop, 23, 130–131
Floorplanning, 17, 81–82, 98, 106–107
FPGA to ASIC conversion, 32–33
Front end, 17, 82, 93, 95
Full-timing-delay model, 143
Functional sign-off, 87
Functional simulation, 25

G

Gate array, 32
Gate-level tools, 144
GDSII, 8, 18, 84
Glitch, 27, 128
Global bus design, 91
Global positioning system (GPS), 135
Global routing, 84–85
Gray coding, 126, 127

H

Hamming distance, 124, 127
HDL, 25, 49
Hierarchical layout block, 89
Hierarchical logic blocks (HLB), 28
Hierarchical methodology, 27, 92
Hierarchical techniques, 100
HW/SW integration, 49
HW/SW partitioning, 121

I

In-place optimization (IPO), 103
Intellectual Property (IP), 1, 7–12
   firm IP, 9
   hard IP, 8
   soft IP, 9
Interface Logic Models (ILM), 27
IP verification, 53–55
IR drop, 85, 103

J

JTAG standard, 37–38

K

Karnaugh map, 136

L

Latch, 23, 129, 130
Layout versus Schematic (LVS), 87
Leakage Current, 115
Lint tools, 25
Logical hierarchy, 99
Low-power design, 111, 119
Low-power design flow, 119
Low-power techniques, 117

M

Makefiles, 54
Memory partitioning, 133
Memory optimization, 132–133
Mesh-based clocking, 84
Methodology, 94
MicroNetwork, 14
Modeling methodology, 73
Modern physical design, 93
MPEG, 2, 5, 65, 73
MPEG Processor, 66
Multiple supply voltage, 122

N

New Design Flow, 96
NMOS, 87

O

On-chip bus (OCB), 44
On-chip interconnect, 6
On-chip processor, 1
Open Core Protocol (OCP), 15, 71, 159–164
Optimization techniques, 85
    architecture-level, 122
    algorithm-level, 121
    RT-level, 124
    gate-level, 136

P

Partial-bus-invert (PBI) coding, 127
Partitioning, 108
Physical compiler, 27
Physical design, 6, 18–19, 81–82
    physical design flow, 82
    modern physical design, 93
Physical hierarchy, 89
Physical prototype, 96
Physical sign-off, 87
Physical synthesis, 102
Physical verification, 86
Pin Grid Array (PGA), 22
Pin assignment rules, 23
Place and route, 18
    Multiple place and route, 92
Placement-based synthesis, 27, 87
PLL, 23, 165–171
    PLL basics, 165
    PLL ideal behavior, 166
    PLL errors, 168
PMOS, 87
Post-routing, 85
Power analysis, 26
Power consumption, 22, 99
Power dissipation, 112
Power estimation, 138–140
Power grid design, 103
Power management, 133–136
Power on-off, 23
Power optimization, 118
    algorithm-level, 121
    architecture-level, 122
    clock gating, 124
    de-glitching, 128
    memory, 132–133
    RT-level, 124
    signal encoding, 124
    gate-level, 136
Power-optimization tools, 140
    behavior-level, 142
    gate-level, 144
    transistor-level, 144
Power supply (V_{DD}), 115
Power estimation, 26
Power reduction, 26
Pulse code modulation (PCM), 5

Q

Quad flat pack (QFP), 22

R

RC extraction, 27
RCS, 54
Regression planning, 49
Resource sharing, 60
Reuse methodology, 19
RISC, 78
Round-Robin Arbitration, 76
RTL coding, 25

S

Set-Top Box (STB), 5, 56, 57
STB characteristics, 58
Short-circuit current, 116
Short-circuit power dissipation, 113, 115
Signal integrity, 85–86
Silicon Ensemble, 86
Silicon virtual prototype, 94, 97, 104
SiliconBackplane, 14, 45, 56, 60
Simulation-based power estimation, 138–140
Sleep mode, 133
Small outline package (SOP), 22
SOC, 1
SOC verification, 47–52
SPICE, 37–38
STAMP models, 27
Standard delay format (SDF), 17, 100
Standard-cell technologies, 32
Static power dissipation, 113, 115
Static timing analysis (STA), 16, 27, 50
  system mode, 30
  test mode, 30
Supply voltage, 22
Surface-mount, 22
Switched capacitance, 117
Switching activity, 117, 123
Synthesis, 25
  logic synthesis, 25
  placement-based synthesis, 27, 87
System partitioning, 63

T

Tapeout, 18
Test Access Port (TAP), 37–38
Testbenches
  BFM-based, 35
  C-based, 35
  vector-based, 35
Time-to-market (TTM), 19, 47, 111
Timing analysis, 6, 27
Timing closure, 93–94, 98
Timing Optimization, 85, 86
Timing sign-off, 87
Transistor-level tools, 144
Turn Around Time (TAT), 12
  degree of difficulty, 24
Two's complement, 123

Index

U

Unit-delay model, 143
Utopia, 4, 6, 67

V

Verification, 104, 109
   analog mixed-signal (AMS), 36–37
   assertion-based verification (ABV), 36
   code coverage, 36
   emulation, 37
   formal verification, 36
   functional verification, 35
   simulation, 35
   testbenches, 35
Verification planning, 48
Verilog, 9, 25, 32, 36, 49, 53, 117, 144
Version control, 54
VHDL, 9, 25, 32, 36, 49, 50, 53, 144

Virtual component (VC), 7
Virtual component interface (VCI), 15
Voice over IP (VoIP), 1, 44, 45, 58
Voltage drop, 85
Voice over Network (VoN), 121

W

Wire-load models (WLM), 25, 96

X

Xilinx Netlist Format (XNF), 32

Z

Zero-delay model, 143

Wouldn't it be great

if the world's leading technical publishers joined forces to deliver their best tech books in a common digital reference platform?

They have. Introducing **InformIT Online Books powered by Safari.**

- **Specific answers to specific questions.**
InformIT Online Books' powerful search engine gives you relevance-ranked results in a matter of seconds.

- **Immediate results.**
With InformIt Online Books, you can select the book you want and view the chapter or section you need immediately.

- **Cut, paste and annotate.**
Paste code to save time and eliminate typographical errors. Make notes on the material you find useful and choose whether or not to share them with your work group.

- **Customized for your enterprise.**
Customize a library for you, your department or your entire organization. You only pay for what you need.

Get your first 14 days FREE!

InformIT Online Books is offering its members a 10 book subscription risk-free for 14 days. Visit **http://www.informit.com/onlinebooks** for details.

informIT

www.informit.com

YOUR GUIDE TO IT REFERENCE

Articles

Keep your edge with thousands of free articles, in-depth features, interviews, and IT reference recommendations — all written by experts you know and trust.

Online Books

Answers in an instant from **InformIT Online Book's** 600+ fully searchable on line books. Sign up now and get your first 14 days **free**.

POWERED BY

Catalog

Review online sample chapters, author biographies and customer rankings and choose exactly the right book from a selection of over 5,000 titles.

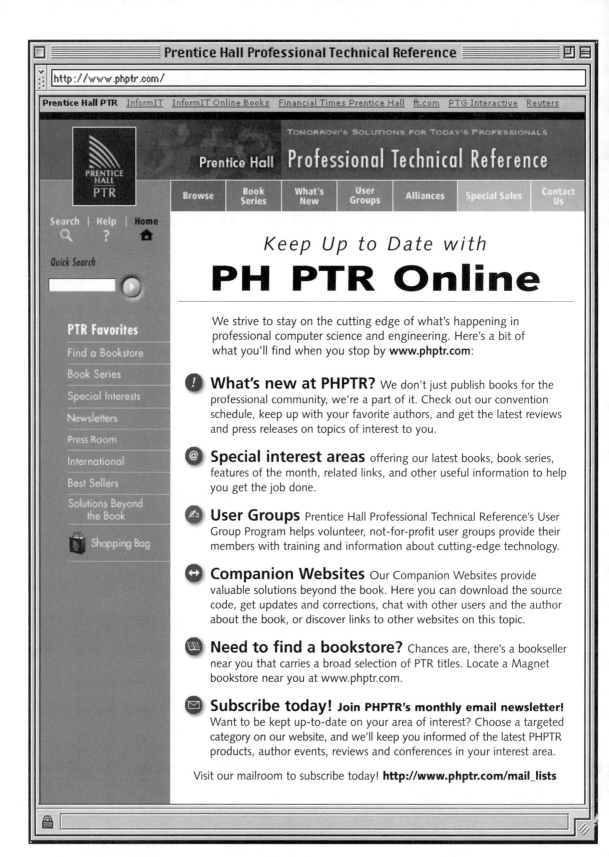